Fog Computing in the Internet of Things

Amir M. Rahmani • Pasi Liljeberg
Jürgo-Sören Preden • Axel Jantsch
Editors

Fog Computing in the Internet of Things

Intelligence at the Edge

Springer

Editors
Amir M. Rahmani
Department of Computer Science
University of California
Irvine, CA, USA

and

Institute of Computer Technology
TU Wien
Vienna, Austria

Jürgo-Sören Preden
Thinnect, Inc., Sunnyvale
CA, USA

Pasi Liljeberg
Department of Future Technologies
University of Turku
Turku, Finland

Axel Jantsch
Institute of Computer Technology
TU Wien
Vienna, Austria

ISBN 978-3-319-57638-1 ISBN 978-3-319-57639-8 (eBook)
DOI 10.1007/978-3-319-57639-8

Library of Congress Control Number: 2017941641

Printed on acid-free paper

This Springer imprint is published by Springer Nature
The registered company is Springer International Publishing AG
The registered company address is: Gewerbestrasse 11, 6330 Cham, Switzerland

Contents

Part I
Introduction on Fog Computing

Chapter 1
Fog Computing Fundamentals in the Internet-of-Things

Behailu Negash, Amir M. Rahmani, Pasi Liljeberg, and Axel Jantsch

1.1 Introduction

The Internet-of-Things (IoT) is a self-configuring and adaptive network which connects real-world things to the Internet enabling them to communicate with other connected objects leading to the realization of a new range of ubiquitous services [1]. This definition of IoT is not comprehensive. There is a variety of definitions that differ in detail as reviewed in [1]. The term *IoT* originates to Massachusetts Institute of Technology Auto-ID center when it was chosen by Kevin Ashton in 1999 [2]. However, the concept of connecting devices to the Internet to remotely monitor their status has been introduced for the first time in 1982 by a group of students at Carnegie Mellon University when they managed to connect a coke machine to the Internet and remotely check its status [3]. Advancements in science and technology enabled making smaller, cheaper, and faster computing devices capable of sensing the environment, communicating and actuating remotely, which resulted to the increased interest of applying the IoT to vast aspects of life, such as smart cities, healthcare, and smart home. Some of these applications are

B. Negash (✉) • P. Liljeberg
University of Turku, Turku, Finland
e-mail: behneg@utu.fi; pakrli@utu.fi

A.M. Rahmani
Department of Computer Science, University of California, Irvine, CA, USA

Institute of Computer Technology, TU Wien, Vienna, Austria
e-mail: amirr1@uci.edu

A. Jantsch
Institute of Computer Technology, TU Wien, Vienna, Austria
e-mail: axel.jantsch@tuwien.ac.at

A.M. Rahmani et al. (eds.), *Fog Computing in the Internet of Things*,
DOI 10.1007/978-3-319-57639-8_1

discussed in detail in the last part of this book in the case study section focusing on the benefits of Fog computing in the respective domains.

The IoT is already around us connecting wearable devices, smart cars, and smart home systems. It is expected that more than 50 billion devices will be connected to the Internet by 2020 [4]. The introduction of such a huge number of connected devices requires a scalable architecture to accommodate them without any degradation of the quality of service demanded by applications. In addition, the majority of the devices that make up the Internet-of-Things are resource-constrained; resources, such as computing power, energy, bandwidth, and storage, are scarce. These constraints limit the deployment scenarios of applications using such IoT devices. For instance, it is infeasible to use a battery-powered sensor to directly connect to the Internet and publish information regarding its surrounding for a long time or store readings of a longer time in local memory. These constraints present a design challenge that is shaping the architecture of the IoT in many ways. Some of these IoT challenges and corresponding efforts in each area are summarized in [5]. Many of these challenges can be mitigated by extending the functions of Cloud computing closer to the IoT devices. Fog computing [6], also known as edge computing, is such an intermediate layer extending the Cloud layer.

Fog computing layer brings computing, network, and storage services closer to the end-nodes in IoT. Compared to Cloud computing, this computing layer is highly distributed and introduces additional services to end-devices located in the perception layer [7]. This bridging layer is referred differently but with similar or small variations in purpose. For example, edge computing [8, 9], micro-cloud [10], or cloudlet [11] are some of the related terms. Regardless of the name, the concept of introducing an intermediate computing layer in IoT is motivated by the similar set of challenges. Moreover, the possible set of services that can be potentially integrated in the Fog computing layer are vast. Some of these services are a scaled version of the ones provided by cloud computing while most of these services emerged recently in response to IoT challenges. This chapter goes through the basics of the requirements of IoT which motivate the benefits of Fog computing and discusses some features, organization, and functions of this layer. Subsequent chapters explain these concepts in more detail and give specific implementation scenarios and use cases of Fog computing. To show the overall organization of the book, we briefly discuss the contribution of each chapter at the end of this introductory chapter.

1.2 Background and Motivation of Fog Computing

The architecture of the IoT is an active research area. Architecture plays a critical role in determining the success of a system. As such, there are several efforts ranging from public projects to industrial standard associations and academic institutions to set a working IoT architecture. Some of these efforts are presented in [12]. Most of the architecture proposals are generic and model IoT regardless of the specific application domain or implementation. The most notable project,

Internet-of-Things Architecture (IoT-A) [13] provides a generalization of IoT domain model that serves as basis for a reference architecture. In the functional view of the model, IoT-A identifies components of an IoT system, for instance, communication, security, management, and IoT services as the main elements. Another generalized component view of IoT systems is presented in [5] where Al-Fuqaha et al. discuss IoT as an integration of identification, sensing, communication, computation, services, and semantics. These modular classifications of IoT are based on the functionality of each unit. Some of these units can be located in a single device. However, an IoT system is naturally a distributed system by definition. Hence, the components identified above are geographically distributed where the communication component is in charge of connecting them. In the simplest form, two groups can be formed: the first group contains identification and sensing while the second one hosts computation, services, and semantics (similar separation can be also achieved in case of IoT-A model). In an effort to find the best level of functional categorization and physical separation, researchers have proposed different alternatives.

A straightforward way to make an IoT device visible through the Internet is to provide it with an access to a Cloud server, such that it can upload data, receive notifications or commands. In such configuration, the client handles reading data from the environment and most of the remaining functions run in the Cloud. This traditional client–server approach of organizing the different components of IoT is still used by many vendors. There are also many variations of this architecture to separate certain logical components of the system into three or five layers [5] (Fig. 1.1). These separations of concerns are mostly based on the functionality of the module. In the three-layer architecture, the sensors appear in the lower perception layer. Network layer, which is located on top of perception layer, connects the sensors to the topmost application layer. The functionality of each layer is distinct in this approach. The sensors and actuators in the perception layer gather the data that will be transported through the network to ultimately reach the application logic. Figure 1.1 shows the different types of logical separation of IoT elements. Other alternatives of this architecture proposal divide the layers into five. Some of these

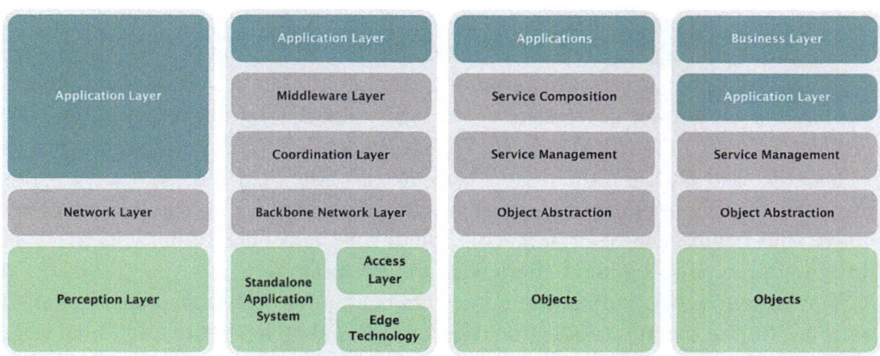

Fig. 1.1 IoT architecture proposals (three layers and five layers) [5]

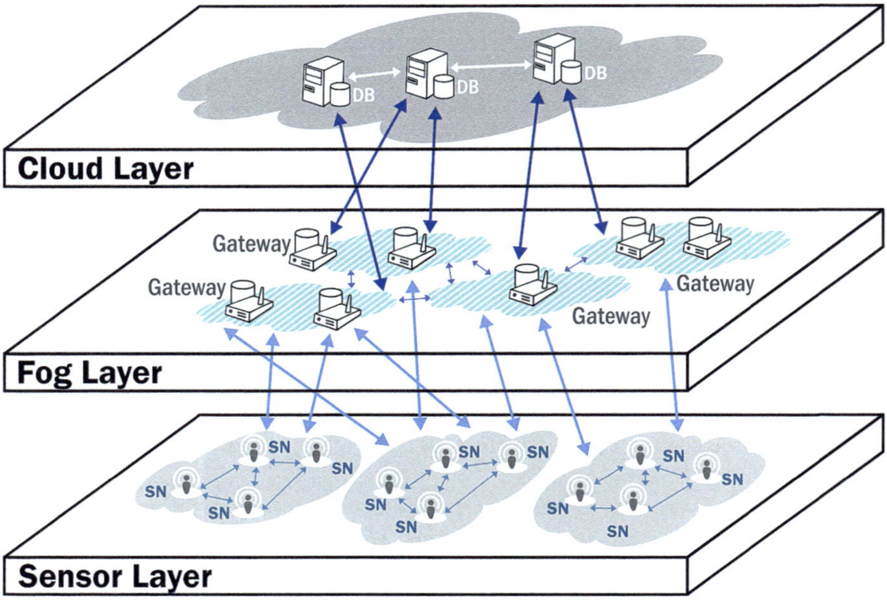

Fig. 1.2 High-level overview of fog based IoT

variations consider middleware and object abstraction as separate layers. These additional layers help provide integration services and encapsulate the devices in the perception layer, respectively. Even though, implementing these layers of logically separated components as such provides modularity and ease of implementation, it fails to address the requirements of the perception layer, such as low latency communication and mobility.

The perception layer or sensor layer, shown in Fig. 1.2, can be composed of millions of devices. The majority of these devices are very tiny in size, battery-powered, and have small memory and limited processing power. Such resource constraints necessitate novel design approaches to accommodate them. In addition, various wireless communication protocols are widely used for networking such as Wi-Fi, Bluetooth Low Energy (BLE), NFC, ZigBee, RFID, and 6LoWPAN. Besides the foregoing network protocol variations, there are differences in application layer protocols even among devices using the same underlying network protocol. For instance, CoAP [14], MQTT [15], DDS [16], and XMPP [17] are among the frequently used ones. Furthermore, there are multiple data formats used by these protocols that are application domain specific. The resource constraints mentioned above, the heterogeneity of protocols, platforms, and data formats, call for the design of more efficient and IoT-friendly architectures.

The design process of a concrete architecture for a system depends on the attributes of the specific application. However, based on the generic IoT challenges and requirements highlighted above, it is possible to have a reasonable generic architecture design. Following on the logical separation of functional components that resulted in three or five layers, we can map the logical components to physical computing layers. As mentioned earlier, in a client–server approach, most of the components (shown in Fig. 1.1) would run in the server located in a cloud. Unfortunately, this approach does not address all the requirements discussed above. This initiated the research for an alternative computing hierarchy that works well for IoT. Fog computing is introduced as an intermediate layer between the perception layer and the Cloud, giving more flexibility of choice for deploying the components of an IoT system architecture. Figure 1.2 shows how Fog computing fits between the perception layer or sensors and the Cloud layer. In the following sections, this layer is discussed by giving more details on the internal organization and services provided.

1.3 Fog Computing Basics

The introduction of the IoT brings billions of devices to the Internet and the majority of these devices are resource-constrained. To overcome the challenges of these devices and meet the requirements of the application domain, the demand for an intermediate computing layer becomes evident. The concept of fog computing is the latest descendant in the line of physical separation of functional units. It is a computing layer closer to the perception layer, where the sensors and actuators reside, and provides computing, networking, and storage services. To accommodate these services, and address the requirements of IoT systems, the Fog layer offers the characteristics discussed in the following.

1.3.1 Characteristics of the Fog Layer

As an intermediate computing layer, the characteristics of the Fog layer are discussed in comparison to the perception and Cloud layers. In contrast to the Cloud layer, the Fog layer is closer to the perception layer and this proximity provides a range of advantages that characterize the layer. One of the immediate benefits over the Cloud is its location-awareness. Such awareness comes due to the large-scale geographical distribution of the devices that make up the Fog layer [6]. As shown in Fig. 1.2, each gateway in the Fog layer manages a subset of nodes in the perception or sensor layer. This subset of resource-constrained devices are located close to each other and the managing gateway can easily locate each device. The location-awareness of the Fog layer can be utilized to address multiple functional and nonfunctional requirements of IoT applications, such as mobility and security. Another closely related characteristic of the Fog layer is its large-scale distribution in contrast to the centralized Cloud layer. Centralization in this context is relative; the Cloud layer is centralized as seen from the client side. Looking from the organization of the servers in the cloud, however, it is geographically distributed but not at the scale expected from the Fog layer. For instance, Cloud service providers such as Amazon have multiple data centers in different regions. The case of geographical distribution in the Fog layer is different due to the small separation distance of the gateways and its large deployment.

The combined benefits of location-awareness and large-scale geographic distribution support the mobility requirements of devices or "things" at the perception layer. An overview of the services available to provide mobility is discussed in the following sections and its practical applications are presented in the later chapters. Moreover, the close-proximity of the Fog layer to the nodes provides real-time interaction mode with the sensors and actuators in the perception layer. The geographical distribution of the Fog layer and subsequently the offered low communication latency are among the critical features of fog layer. Some IoT application domains, such as healthcare or automotive, are highly dependent on such feature. For instance, a case study of edge-based ECG feature extraction is presented in [18].

The IoT in general is dominated by wireless networks. There are many wireless protocols, mostly tailored for low-power operation, coverage, or bandwidth. For instance, 6LoWPAN [19], BLE, NarrowBand IoT Protocol (NB-IoT) [20], LoRa [21], and Sigfox [22] are some of these protocols. The majority of these protocols connect sensor nodes to the Fog layer to get access to the Internet. These protocols are usually incompatible with each other. To cope with this issue, Fog layer provides an additional benefit of acting as an interpretability layer among these heterogeneous protocols. There are several middleware proposals that utilize this layer as a means to translate or adapt different network or application protocols [23]. The gateways in the Fog layer can also perform lightweight analytics at the edge to give feedback, command, and notification to the end-users as well as the sensor nodes in real-time. In addition, the internal organization of the Fog itself can be arranged in a federated or hierarchical way based on functional or location of the connected devices.

1.3.2 Design and Organization of the Fog Layer

Based on the characteristics of the Fog layer presented in the previous section and the possible set of services highlighted in the following section, the Fog layer can be organized in an efficient way to address the requirements. This section is by no means comprehensive and detailed enough to build a usable intermediate layer, but instead gives an introductory information that will be dealt in more detail in later chapters. To begin with, consider a network gateway or a wireless hot spot serving clients in its vicinity. The role of such a gateway is to pass network packets to the back-end infrastructure which is connected to the Internet. In a larger environment, multiple access points can be arranged to provide users with seamless connectivity throughout the intended area. Considering the tons of devices that connect to the Fog layer, this layer can be visualized as a network of gateways covering a larger area. In addition to simply passing network packets, these networked smart gateways can process the data or store it when necessary [18]. Figure 1.2 shows the Fog layer where distributed smart gateways communicate with the Cloud, the sensor layer, and among themselves. In the gateways of the Fog layer, the network interface is a critical component to enable support for the various wireless network protocols shown in Sect. 1.3.1.

1.4 Fog Computing Services

The characteristics of Fog computing layer have been highlighted in Sect. 1.3.1. We mentioned that these characteristics can be leveraged to provide services that assist the perception layer, such that the overall system requirements are met. This layer takes advantage of its proximity to the sensor layer and provides services that are extensions of the cloud layer and also unique ones that are feasible only at this layer. This section gives an overview of a subset of possible services at the fog layer and the related advantages that enable IoT. These services are organized into Compute, Storage, and Network services.

1.4.1 Computing Services

The limitations of computing power of the devices in the perception layer have led to the introduction of remote processing approaches. Processing at the Fog layer is not only motivated by the constraint of processing power at sensor nodes but also by the desired location of computing to better meet system requirements and maintain energy efficiency. Earlier Cloud-based processing can be brought down to the Fog layer for localized processing and immediate response [24, 25]. In this regard, there can be multiple configurations of sharing the computing load among the different layers in the IoT-based system, and the processing requirements may vary based on the actual work. For instance, considering a system which performs data processing to learn a certain pattern, the workload can be distributed in such a way that localized patterns can be identified in the Fog layer while the generalized patterns are only available at the Cloud. This load sharing is discussed in detail in upcoming chapters. Beside data management, events can be handled at the Fog layer. The proximity of this layer makes it an ideal candidate to handle events to react in real-time and enhance the reliability of the system. Moreover, there are many middleware that leverage the Fog layer to manage physical devices through abstraction, agent-based management, and virtual machines. The following sections provide an introductory overview of these computing services that can be realized at the Fog layer.

1.4.2 Storage Services

A huge amount of data can be generated by the sensor nodes and there are billions of these sensor devices around. The storage available in the devices at the perception layer is not often sufficient to store even a 1-day data considering the rate of data generation. As discussed earlier, pushing all the data directly to the cloud is not necessary in particular when there is irrelevance or redundancy in data. The wise approach in such cases would be to filter and temporally store the data in the intermediate fog layer [26, 27]. Combined with the computing service, the stored data can be filtered, analyzed, and compressed for efficient transmission or for learning local information regarding the system behavior. In cases where the communication may not be robust, the storage services help enhance the reliability of the system by maintaining proper system behavior for client nodes. Sarkar et al. [28] present such features of fog layer in their assessment of the fog computing for IoT.

1.4.3 Communication Services

The communication in the IoT is dominated by wireless nodes. Due to the resource constraints in the perception layer, these wireless protocols are optimized for low-power operation, narrow-band transmission or longer range of coverage. Currently, a long list of alternative protocols are available in the market [29]. The Fog layer is located in a strategic place to organize these multitude of wireless protocols and unify their communication to the Cloud layer. This helps in managing subnetworks of sensors and actuators providing security, channeling messages among devices, and enhancing the reliability of the system. In addition, this layer can provide interoperability of disparate protocols by listing and interpreting the representation format. Moreover, the Fog layer provides visibility of devices that are non-IP-based to be accessible through the Internet [26].

1.5 Summary and the Book Organization

This chapter presented a concise introduction of Fog computing in IoT. It covered the driving reasons for the design of IoT systems reinforced with the fog computing as an intermediate computing layer. In addition, a high-level introduction of the internal structure and organization of the layer was discussed. As a motivation to the demand for Fog computing, we showed the functional units of an IoT system and distribution of these units through physical computing layers. In doing so, however, IoT system requirements such as mobility and resource constraints of the perception layer demand a nearby layer. This layer, introduced as Fog due to its proximity

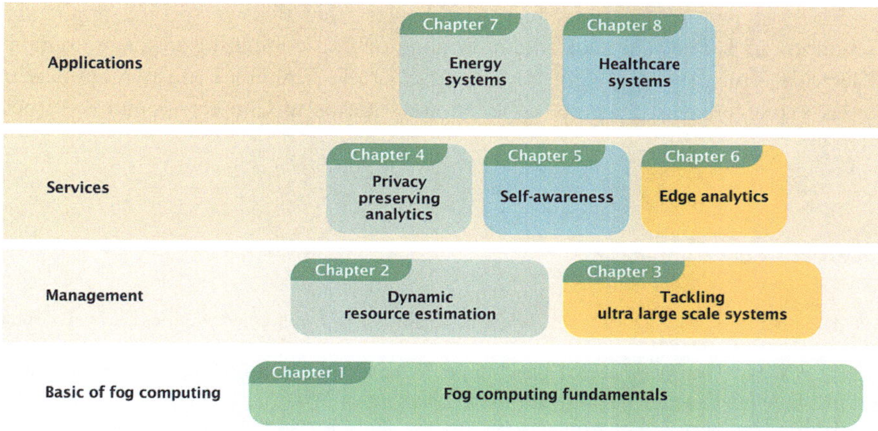

Fig. 1.3 Organization of the chapters

to the ground compared to the Cloud in its literal meaning, provides connectivity, storage, and processing for sensors and actuators. To achieve these functions and scale, the Fog layer has modular organization and is geographically distributed. These provided services via Fog Computing are classified into three main classes as compute, storage, and network functions supporting the sensor nodes.

The heterogeneous nature of the wireless network protocols, platforms, and architectures in the perception layer of IoT makes it difficult to build an integrated and reliable system. The Fog layer provides services that can be used to hide such heterogeneity and provide a uniform access channel to the perception layer for users through the Internet. These concepts are well examined with practical implementations and evaluation of certain aspects of their performance in upcoming chapters.

This book is organized into eight chapters to provide readers with a comprehensive information about fog computing in the context of IoT. Chapter 1 is a preliminary overview of what fog computing is, its characteristics, and the possible set of services that can be hosted in this layer to meet system requirements of IoT. In essence, it gives a general background that serves as foundation to understand subsequent chapters. Figure 1.3 shows the organization of the book in general. This chapter gives wider but shallower conceptual basis that is enforced with two chapters going into details of the internal structure of Fog computing focusing on management. Chapters 2 and 3 give details on Fog based IoT scalability and resource estimation of the Fog layer to accommodate the billions of additional Internet connected devices. The third part of the book contains three chapters that go further in discussing some of the critical services needed at the Fog layer. Chapters 4, 5, and 6 focus on security and privacy of the IoT as realized through the Fog layer, learning and self-awareness of IoT systems through Fog computing, and detailed data analytics in a smart city application.

The last part of the book has two chapters that discuss specific application scenarios of IoT systems and the advantage of Fog computing in those domains. Electrical grid control system implementation and healthcare are two application areas explained with concrete application scenarios in Chapters 7 and 8, respectively.

References

1. R. Minerva, A. Biru, D. Rotondi, Towards a definition of the internet of things (IoT). Technical report, IEEE, 2015
2. K. Ashton, That 'Internet of Things' thing: In the real world, things matter more than ideas, RFID Journal, http://www.rfidjournal.com/articles/view?4986
3. CMU, The "only" coke machine on the internet, http://www.cs.cmu.edu/~coke/
4. S. Ray, Y. Jin, A. Raychowdhury, The changing computing paradigm with internet of things: a tutorial introduction. IEEE Des. Test **33**(2), 76–96 (2016)
5. A. Al-Fuqaha, M. Guizani, M. Mohammadi, M. Aledhari, M. Ayyash, Internet of things: a survey on enabling technologies, protocols, and applications. IEEE Commun. Surv. Tutorials **17**(4), 2347–2376, Fourthquarter (2015)
6. F. Bonomi, R. Milito, J. Zhu, S. Addepalli, Fog computing and its role in the internet of things, in *Proceedings of the First Edition of the MCC Workshop on Mobile Cloud Computing*, MCC '12 (ACM, New York, 2012), pp. 13–16
7. S. Yi, C. Li, Q. Li, A survey of fog computing: concepts, applications and issues, in *Proceedings of the 2015 Workshop on Mobile Big Data*, Mobidata '15 (ACM, New York, 2015), pp. 37–42
8. F. Jalali, A. Vishwanath, J. de Hoog, F. Suits, Interconnecting fog computing and microgrids for greening IoT, in *2016 IEEE Innovative Smart Grid Technologies – Asia (ISGT-Asia)*, Nov 2016, pp. 693–698
9. O. Salman, I. Elhajj, A. Kayssi, A. Chehab, Edge computing enabling the internet of things, in *2015 IEEE 2nd World Forum on Internet of Things (WF-IoT)*, Dec 2015, pp. 603–608
10. M. Selimi, L. Cerdà-Alabern, L. Wang, A. Sathiaseelan, L. Veiga, F. Freitag, Bandwidth-aware service placement in community network micro-clouds, in *2016 IEEE 41st Conference on Local Computer Networks (LCN)*, Nov 2016, pp. 220–223
11. M. Satyanarayanan, P. Simoens, Y. Xiao, P. Pillai, Z. Chen, K. Ha, W. Hu, B. Amos, Edge analytics in the internet of things. IEEE Pervasive Comput. **14**(2), 24–31 (2015)
12. S. Krco, B. Pokric, F. Carrez, Designing IoT architecture(s): a European perspective, in *2014 IEEE World Forum on Internet of Things (WF-IoT)*, Mar 2014, pp. 79–84
13. IoT-A Project, Internet of things – architecture, IoT-A, deliverable d1.5 – final architecture reference model for the IoT v3.0. Technical report, EU-FP7, 2013
14. Z. Shelby, K. Hartke, C. Bormann, The constrained application protocol (CoAP), https://tools.ietf.org/html/rfc7252
15. OASIS, MQTT version 3.1.1 oasis standard, http://docs.oasis-open.org/mqtt/mqtt/v3.1.1/os/mqtt-v3.1.1-os.pdf
16. OMG, Data distribution service DDS, http://www.omg.org/spec/DDS/1.4/
17. XSF, Extensible messaging and presence protocol (XMPP), https://xmpp.org/extensions/index.html
18. T.N. Gia, M. Jiang, A.M. Rahmani, T. Westerlund, P. Liljeberg, H. Tenhunen, Fog computing in healthcare internet of things: a case study on ECG feature extraction, in *2015 IEEE International Conference on Computer and Information Technology; Ubiquitous Computing and Communications; Dependable, Autonomic and Secure Computing; Pervasive Intelligence and Computing*, Oct 2015, pp. 356–363

19. N. Kushalnagar, G. Montenegro, C. Schumacher, Ipv6 over low-power wireless personal area networks (6LoWPANs): overview, assumptions, problem statement, and goals, https://tools.ietf.org/html/rfc4919
20. Y.-P.E. Wang, X. Lin, A. Adhikary, A. Grövlen, Y. Sui, Y.W. Blankenship, J. Bergman, H. Shokri-Razaghi, A primer on 3GPPnarrowband internet of things (NB-IoT). CoRR, abs/1606.04171 (2016)
21. LoRa Alliance, LoRa wide area network for IoT, https://www.lora-alliance.org/What-Is-LoRa/Technology
22. Sigfox, About sigfox, http://www.sigfox.com/
23. B. Negash, A.M. Rahmani, T. Westerlund, P. Liljeberg, H. Tenhunen, LISA 2.0: lightweight internet of things service bus architecture using node centric networking. J. Ambient Intell. Humaniz. Comput. **7**(3), 305–319 (2016)
24. S.K. Datta, C. Bonnet, J. Haerri, Fog computing architecture to enable consumer centric internet of things services, in *2015 International Symposium on Consumer Electronics (ISCE)*, June 2015, pp. 1–2
25. P. Hu, H. Ning, T. Qiu, Y. Zhang, X. Luo, Fog computing-based face identification and resolution scheme in internet of things. IEEE Trans. Ind. Inf. **PP**(99), 1–1 (2016)
26. A.M. Rahmani, N.K. Thanigaivelan, T.N. Gia, J. Granados, B. Negash, P. Liljeberg, H. Tenhunen, Smart e-health gateway: bringing intelligence to internet-of-things based ubiquitous healthcare systems, in *2015 12th Annual IEEE Consumer Communications and Networking Conference (CCNC)*, Jan 2015, pp. 826–834
27. A.M. Rahmani, T.N. Gia, B. Negash, A. Anzanpour, I. Azimi, M. Jiang, P. Liljeberg, Exploiting smart e-health gateways at the edge of healthcare internet-of-things: a fog computing approach. Futur. Gener. Comput. Syst. (2017). http://www.sciencedirect.com/science/article/pii/S0167739X17302121
28. S. Sarkar, S. Chatterjee, S. Misra, Assessment of the suitability of fog computing in the context of internet of things. IEEE Trans. Cloud Comput. **PP**(99), 1–1 (2015)
29. Z. Sheng, S. Yang, Y. Yu, A.V. Vasilakos, J.A. Mccann, K.K. Leung, A survey on the ietf protocol suite for the internet of things: standards, challenges, and opportunities. IEEE Wirel. Commun. **20**(6), 91–98 (2013)

Part II
Management at the Fog Layer

Chapter 2
IoT Resource Estimation Challenges and Modeling in Fog

Mohammad Aazam, Marc St-Hilaire, Chung-Horng Lung, Ioannis Lambadaris, and Eui-Nam Huh

2.1 Introduction

With the advancements in sensor-networking, gadgets, and digital devices, services encompassing Internet of Things (IoT) are rapidly gaining popularity. As of now, there are around ten billion connected devices already, expected to become 24 billion by 2020 [1]. Cisco and Ericsson even made a prediction of 50 billion devices by the end of this decade [2, 3]. With such a trend, a lot of devices would be there to become part of IoT. Those devices as well as the data would be heterogeneous as well, generated in irregular frequency. In such a case, IoT on its own would not be self-sufficient to manage the challenges associated [4]. A lot of tasks would be handled through fog computing. Additionally, several services [5] would be provided by a cloud as well because IoT and sensors would be resource-constrained. In that case, fog would be a mediator and will be able to perform tasks that may not be efficiently done by a distant cloud. Fog would be present close to the underlying nodes for the purpose of offloading the tasks and preprocessing the raw data. Fog will also be able to minimize delays and enhance service quality by incorporating better responsiveness. Hence, it would be inevitable for multimedia streaming and other delay sensitive services to operate without a fog. Scenarios where multimedia services are to be provided, like Video on Demand (VoD), Visual Sensor Network, or CCTV connected to cloud, etc. consume

M. Aazam (✉) • M. St-Hilaire • C.-H. Lung • I. Lambadaris
Department of Systems and Computer Engineering, Carleton University, Ottawa, ON, K1S 5B6, Canada
e-mail: aazam@ieee.org; marc_st_hilaire@carleton.ca; chlung@sce.carleton.ca; ioannis@sce.carleton.ca

E.-N. Huh
Department of Computer Engineering, Kyung Hee University, Suwon, Republic of Korea
e-mail: johnhuh@khu.ac.kr

A.M. Rahmani et al. (eds.), *Fog Computing in the Internet of Things*,
DOI 10.1007/978-3-319-57639-8_2

more processing power and storage space. Resource scheduling becomes tricky, especially in the case of mobile nodes. Resource underutilization becomes more of a problem when underlying nodes are mobile. Service provider has to deal with unexpected and unreliable service utilization behavior of its customers, henceforth, requiring for dynamic and customer's historical record-based resource allocation. Fog can play an important role in this case, being close to the proximity of the users and able to make decisions dynamically in a more realistic way. Furthermore, mission critical and latency sensitive IoT services require very quick response and processing. In that case, it is not feasible to communicate through the distant cloud that is accessible over the Internet. A CoT-like scenario would be suitable in such cases, where offloading and latency-sensitive tasks are dealt with locally through fog computing, while deep learning, big data, long-term storage, and other processing rich tasks are done by the cloud [5].

The concept of fog computing is to bring networking resources near the nodes that are generating data. We refer to it as perception layer since data is to be perceived from there. Fog creates another layer on top of perception layer, while cloud resides above all in the cloud layer. Fog resources lie between the perception layer and the cloud layer. Fog computing is an extension of the traditional cloud computing paradigm to the edge of the network, helping to create more refined and context-aware services [6]. For mobile nodes, like moving vehicles or drones, fog provides low latency and high quality streaming through proxies and access points located accordingly along highways and tracks. Likewise, resource and power constrained individual nodes, WSNs, and Virtual Sensor Networks (VSNs) would be able to take advantage from the presence of fogs. Fog also suits services related emergency and disaster management, gaming, healthcare, augmented reality, graph/data mining, etc. [7]. But it would hugely depend on how resources are managed in fog computing. Resource management must be very dynamic and in accordance with the type of service and type of devices. Communication means also play an important role, since being mobile or static means that different resource allocation is required even for the same service. One of the ways to dynamically estimate resources is to incorporate the usage pattern and history of the customers of a service. As history will give predictability to the fog, tailored resource estimation can be performed.

This chapter is based on our previous work presented with more details in [8]. We provide a methodology for historical record-based resource estimation to mitigate resource underutilization as well as enhance service quality for multimedia IoTs. The contributions of this work are twofolds. First, we extend the traditional resource management of IoTs at fog by incorporating historical records of cloud service customer (CSC), which can help in efficient, effective, and fair management of resources. Second, to enhance QoS and QoE, the resources are estimated on the basis of the type of the requesting device. The model is implemented using Java and evaluated in cloud-fog scenario using CloudSim. We also discuss some of the main challenging scenarios where fog resource management requires unconventional methods of resource estimation.

2.2 Contributions of the Book Chapter

This book chapter provides how dynamic resource estimation can be provided through fog, on the basis of relinquish probabilities of the customers and the service quality experienced. In this way, a dynamic resource determination model is introduced that takes into account the stakes of customer as well as the provider. Customer can get what it deserves; on the other hand, provider gets the reliability of its customers, minimizing resource underutilization and chances of service give-up.

2.3 Related Work

Fog computing is still a very new concept due to which there is not a lot of literature available on it. Most of the works still focus mainly on cloud resource management. The scenario of fog computing or CoT has not been deliberated by most of the past works. Below are a few examples of relevant papers on IoT resource management.

Abu-Elkheir et al. [9] elaborate on the organization of data in IoT. The authors indicate how distinctive design parameters would work for management of the data. But how that data and IoT nodes are going to be handled at the cloud layer and how resources are to be managed for the generated data are not part of this study. Cubo et al. [10] present their work on the integration of heterogeneous devices accessible via cloud. The presented work, however, does not deal with the key issue of management of resources for such devices in the cloud. Ning and Wang discuss in [11] the potentials of IoTs and the volume of data it is going to produce. The authors also underscore efficient management of resources for the future Internet, in which heterogeneous IoTs would be a vital part. Chatterjee and Misra [12] provide a mapping of sensors to their respective targets through a sensor-cloud infrastructure. But how every node or sensor is allocated with resources in a dynamic fashion is not part of this study. Sammarco and Iera [13] analyze Constrained Application Protocol (CoAP) for IoTs and discuss service management and method for exploiting resources of IoT nodes. Tei and Gurgen [14] emphasize on the significance of the integration of cloud-IoT. They discuss preliminary outcomes of a project in this domain. In [14], Rakpong et al. consider resource allocation in mobile cloud computing environment. They deliberate on communication/radio resources and computing resources, but their work only focuses on decision-making for coalition of resources to increase service provider's revenue. Distefano et al. [15] contribute in presenting an outline for the integration of the underlying IoT nodes with the cloud. However, the challenge of dynamic and node-based resource management is not a focus of this study. Bonomi et al. [6] present a basic architecture for fog computing which does not include its practical implications and resource management for IoTs. Similarly, Stolfo et al. [16] present data protection through fog computing but do not discuss resource management and associated matters.

2.4 Fog Computing

Fog computing is a recently evolved paradigm, extends traditional cloud close to the underlying nodes. As it extends cloud to the edge of the network, it is also known as edge computing. Compared to traditional cloud, fog is a Micro Data Center (MDC), low in potential and resources as compared to a full-fledged cloud datacenter.

Fog provides computation, storage, and networking services to the end nodes in an IoT [6]. Fog is targeted for widely distributed services because of being close to the underlying nodes. Cloud, on the other hand, is centralized and does not suit such applications. Figure 2.1 presents an overall architecture of fog computing where fog works as a middleware, extending cloud and providing resources to the underlying sensors, home networks, body area and healthcare networks, VANETs, etc.

Fogs also comprise gateway(s) which can help achieve data communications in a smarter way in accordance with the requirement of the higher level applications as well as the constraints of underlying nodes. Such type of gateway is termed as Fog-Smart Gateway (FSG) [5] or simply smart gateway. In the cloud-fog-IoT architecture presented in Fig. 2.1, the underlying nodes and networks may not always be physical. Virtual sensors and VSNs are also required for various services. Several services do not require continuous communication. For example, a CCTV network may not require bandwidth consuming video content to be sent to the cloud during some part of the night. Facial recognition or motion detection algorithms running on fog resources assisted with FSG can decide when and what to communicate, hence, saving scarce bandwidth and storage space. Data can then be temporarily stored at fog. Many other applications also require temporary data storage for which fog will be the best option. Moreover, preprocessing, security and privacy enhancements, data trimming, content delivery services, context-aware services, etc. would be best suited with fog. In addition to that, in future, multiple WSNs and IoTs would be present with heterogeneous devices and protocol support. Fog can play its role in integrating such heterogeneous nodes, creating IoT federation and extending the scope of IoT. This will help in the inception of enhanced and rich services.

2.5 Fog Resource Estimation and Its Challenges

Resource estimation in fog is more of a challenge as compared to standard cloud. Reason being the nature of underlying nodes fog must deal with. Although the devices that access a cloud are also heterogeneous and generate diverse types of data, but since fog is also majorly dealing with IoT nodes, the nodes are way too random. Depending upon the type of IoT service, the underlying nodes can be devices as well as dumb objects. A multitude of mobile nodes, including fast moving vehicles would also be among the pool of nodes requiring resources from a fog. In addition to that, when it comes to deal with sensitive data, an additional security layer is to be involved, which fog is responsible for. All this is where fog's resource

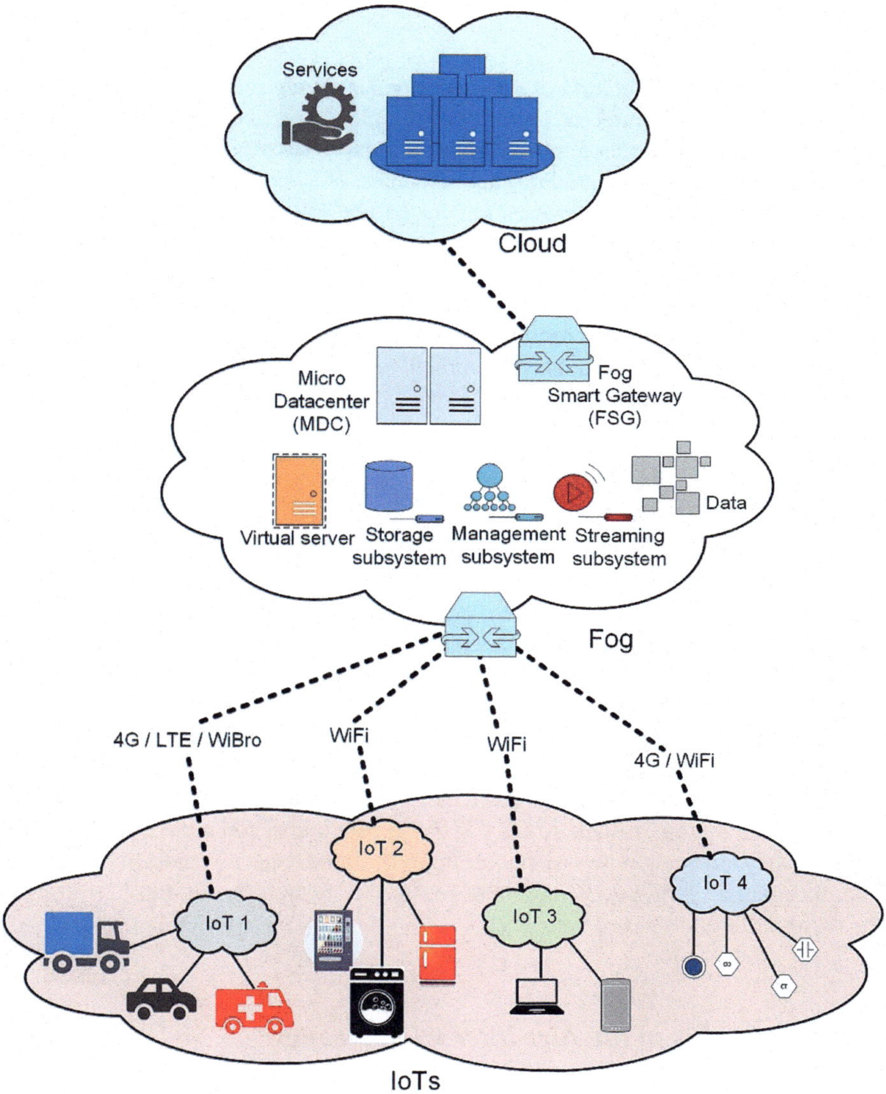

Fig. 2.1 Fog as a middleware between Internet of Things (IoTs) and cloud

management becomes too challenging. Cloud is always accessible over the Internet. Therefore, whatever devices request for cloud resources go through the latency of the core network at least. On the other hand in the case of fog, a more customized and service-oriented resource estimation has to be performed. Following are some of the examples and scenarios where fog resource management is challenged.

2.5.1 Device Type

Fog has to consider what kind of device or node is asking for what kind of resource. A resource rich computer or laptop with ample power usually requires a lot of resources with high service quality expectations. A smartphone will ask for even quicker response, requiring additional resources. However, power may become an issue in many cases; thus, services may have to be offloaded from such device to a fog. A sensor would be resource-constrained; hence, fog has to allocate resources keeping in view the power or battery status of the sensor. Dumb objects connected through RFID and other such means do not require a quick response in most of the cases. However, processing resource would be the key as the data generated from them may have to be accumulated, trimmed, and processed before sending it to the cloud or creating services locally.

2.5.2 Mobility on the Ground

The future of technology involves a lot of sophisticated mobile devices. Smartphones already exist and with the pace the advancements are being made in smartphone technology, they will be no less than a small server very soon. Already a smartphone carries a dozen sensors on an average. With mobility, the resource procurement becomes more of a challenge. Similarly, with the research going on in the domain of VANETs, a lot of private and public vehicles and the whole transport system will become part of the Internet. Several crowdsensing applications will take benefit from fog-based IoT services. Several sensors within a vehicle will be working under a fog. In such a case, fog has to take into account the type of sensors, the way they are being powered, data communication frequency, mobility speed, and mobility pattern while deciding about reasonable resources. Figure 2.2 illustrates this with different types of vehicles.

2.5.3 Mobility in the Air: Internet of Drones

Studies on drones, also known as Unmanned Aerial Vehicles (UAVs), are also getting matured and prototypes are being developed. Amazon and Google have already been working on delivery drones for goods and food. The future of drones would be Internet of Drones (IoD), where multiple drones will be working in tandem. That would be possible only through a fog. For such flying objects, resource estimation would be way beyond the conventions. IoD would require much faster processing and high bandwidth. Some drones would generate high-definition video data, some will be responsible for imagery, and some would be equipped with some other sensors or an array of sensors. Drones would be required to be controlled

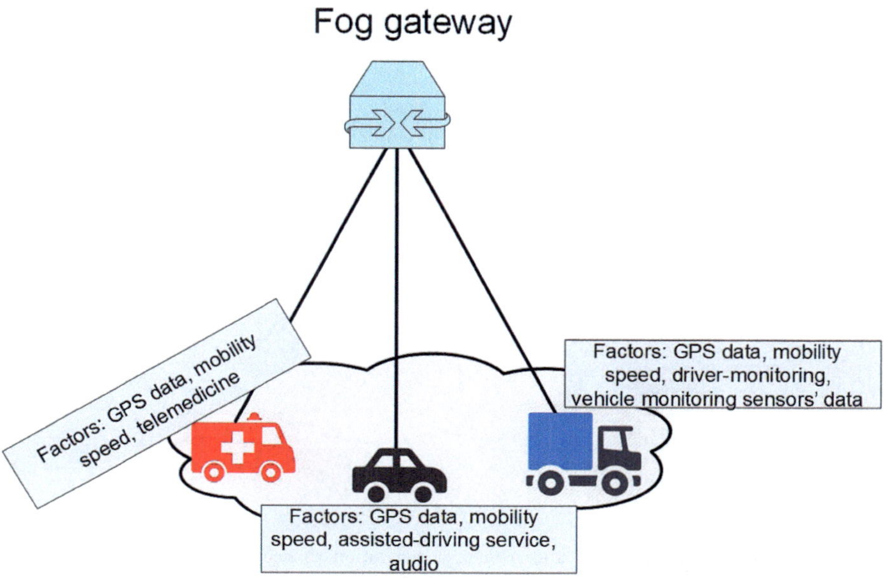

Fig. 2.2 Fog monitoring various factors for resource allocation

from the ground. Communication among the drones would be maneuvered from the ground controller, which essentially would be part of a fog network. For locations where Internet connectivity is challenging, drones can be used as Vehicles of Internet (VoI). In that case, fog network would be consisting of base station drones as Wireless Broadband (WiBro) carriers. After all, fog has to make decision on dynamic and correct resource estimation, keeping in view the behavior of the requesting node and the type of service.

2.5.4 Power Utilization and Status

Although some devices would perform their part of processing locally, many would require it to be done at the fog. Eventually, in many cases, it depends on the status of power on a device and its requirements of amount of power. Fog is responsible for monitoring power and deciding when to offload task(s) from a device. The resource estimation would be performed dynamically in real-time. Figure 2.3 depicts how various factors in power utilization and data type influence fog's decision-making.

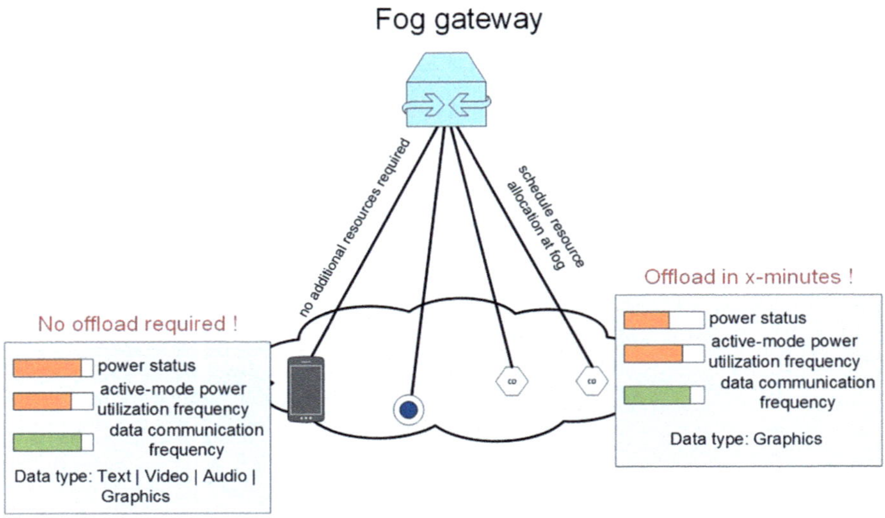

Fig. 2.3 Fog monitoring power and data communication to decide about resource allocation locally

2.5.5 Data Type

Type of data plays a very important role in assessing the type and amount of resources. Multimedia data would require processing, memory, storage, and GPU. On the other hand, text-based data requires the data to be intact (i.e., guaranteed processing). Storage and memory depends on the amount of data and nature of application.

2.5.6 Security

When it comes to sensitive data and applications, like location-based, healthcare, and military, data requires to be concealed before sending to the cloud. This extra layer of security requires more processing; hence, fog has to estimate resources accordingly. Security is essentially of two types here: data security and communication security. Data security refers to making the data unreadable for unintended party. While communication security means the data is transferred through a secure channel.

2.5.7 *Customer's Reliability and Loyalty of Service Utilization*

As it has been discussed that heterogeneous devices are part of IoT with differing data generation patterns and geographical locations, it would be very difficult to forecast if a requesting customer is going to fully utilize the resources requested. Especially with mobile nodes, reliability cannot be guaranteed. If a certain check is set on customer's behavior and service utilization pattern, better resource estimation can be performed. Otherwise, fog will be left with underutilized resources.

2.6 Customer's Reliability-Based Dynamic Resource Estimation in Fog

IoT nodes, sensors, and CSCs contact a fog to acquire the resources they require in accordance with the type of service being utilized. Fog's responsibility is to provide best possible service to the customers, keeping the balance and fairness among all users and stakeholders. For this purpose, fog considers the reliability of customers. Reliability in the sense of service utilization is referred to as Relinquish Rate (RR) here. Based on the previous historical record of each customer, fog estimates resources. For new customer, where history is not yet built, default resource estimation is performed. Hereafter, a reliable customer gets most out of it while an unreliable one will get what it deserves. Although, to encourage a relatively unreliable customer to become more reliable, fog increases the resources as the resource utilization continues. This works as an incentive as well.

We formulate the estimation of required resources as:

$$\mathfrak{R} = \sum\nolimits_{i=0}^{n} \left(\left(U_i{}^* \left(1 - \overline{X} \cdot (P_i(L/H)_s) \right) - \sigma^2 \right)^* (1 - \Omega_i)^* \phi \right) \tag{2.1}$$

$$\mathfrak{R} \in \{CPU, memory, storage, bandwidth\}$$

where $\mathfrak{R}$ represents required resources, and U_i is the base price of the service i in request. Mostly, U_i is determined at the time of contract negotiation. Service-oriented relinquish probability (SR) is represented by $(P_i(L|\, H)_s)$. An average $\overline{\mathrm{x}}\cdot$ is taken for all available SRs for a customer. SR tells the RR for any specific service s. In case the service is requested for the first time, the default value set for $\overline{\mathrm{x}} \cdot (P_i(L|H)_s)$ is 0.3. Generally, probability range is from 0 to 1. In our case, since we are talking about relinquish ratio, 0 would be interpreted as service that has never been relinquished, while 1 would mean that service has always been relinquished completely. In other words, it was never used. For comprehension, we take RR range as greater than 0 and less than 1. Rounding it up and off, respectively, would result in an RR range of 0.1–0.9. Subsequently, the first half 0.1–0.5 represents relatively loyal customer and the second half >0.5–0.9 is to show relatively disloyal customer.

$$0 < L \leq 0.5, 0.5 < H \leq 1 \tag{2.2}$$

$$\Omega_i = \begin{cases} \overline{x} \cdot (\overline{x} \cdot (\sum_{k=0}^{n} P\left(L/H\right)_k)), P(L/H)_{\text{last}}) & \text{if } n > 0, \\ 0.3 & \text{if } n > 0 \end{cases} \tag{2.3}$$

Customers' RR can be misleading in many cases, especially when there are not many request histories been gathered or customer's behavior has been very fluctuating. To counter this issue, we incorporate variance of SR, represented by σ^2. Variance helps determining the actual behavior of each customer.

While SR is explained before as service specific RR, when CSCs use different services, they build an overall relinquish rate (OR) as well. OR is to tell the overall behavior of the customer. Represented by Ω, OR is a decision variable value assigned to a CSC by fog, based on CSC's overall loyalty. Here, it is to be emphasized that $P_i(L|H)_s$ shows probability of the same service that is being asked by the customer currently. While Ω represents overall relinquish probability, which includes all services a customer has used so far. Including all activities a certain customer has been doing with the service provider. Most recent behavior is determined from the last relinquish probability. That is why, it is given more importance and the average is taken again by adding last RR. Same as the case of SR, default value of OR is also set as 0.3, when there is no previous record. In this case, it would mean that CSC has never used any service before with the current service provider.

In our model, device type is deemed crucial while determining resources. Represented by Φ, device type tells fog what kind of device is in use while making the request. Accessing device is only meant to use a specific service, therefore, device capabilities, like CPU and memory, are not currently considered in our model. However, for multimedia services, display size has a role to play, because fog has to allocate resources according to the display size and fitting quality. Device's mobility also comes into play here. A mobile device would require more resources from the fog, because it is in motion and requires quick response for efficient transcoding and data delivery. Laptop can be used in static mode as well as mobile. In either case, different resources would be required for the same kind of service. To determine which mode is a laptop device being used, we consider the access network through which a request has been made. 3G/4G and LTE/LTE-A would be the cases when laptop is used in a mobile mode. In the future extension, this can be done through GPS coordinates as 3G/4G dongles can be used in static mode as well. If laptop is always considered as a mobile device while it is actually being used in static mode, precious datacenter resources would be wasted. Based on our real experiments in different wired and wireless networks (broadband, WiFi, WiBro, 3G, and 4G LTE-A) [17], we came to the conclusion that smartphone and similar devices would require approximately 1.25 of the resources reserved for static device (desktop computer or laptop in static mode). On the other hand, large mobile device (tablet and laptop) requires approximately 1.5 times of such resources. Main focus should be to give priority to service accessed by mobile devices, instead of doubling the resources. Therefore, the value of Φ would accordingly be:

$$\phi = \begin{cases} \phi_{m_s} = 1.25 \\ \phi_{m_l} = 1.5 \\ \phi_s = 1 \end{cases} \tag{2.4}$$

where ϕ_{m_s} represents small mobile device, ϕ_{m_l} represents large mobile device, and ϕ_s represents static device.

With this formulation, fog can determine future resource requirements. Correct resource estimation will also help in managing power consumption which is becoming a point of concern for datacenters.

2.6.1 Resource Estimation for a New Customer with No Previous History

When a request for any service is made by a CSC, fog prioritizes loyal customer while being generous in estimating resources. For very new customers, fog uses default relinquish ratio as there is no history available for them. In other words, it is expected that this new customer will be "fairly" loyal. That is why, relinquish probability is set to 0.3. Even though the minimum value of RR in our model, which happens to be 0.1, could have been set for new customer, but since datacenter resources are precious and it is not advisable to take risk (which can have more impact in case of expensive services or long-term subscriptions), therefore, instead

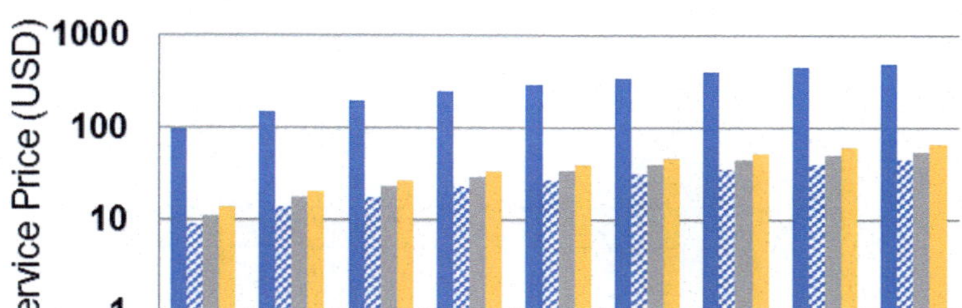

	1	2	3	4	5	6	7	8	9
Provider's Price	100	150	200	250	300	350	400	450	500
Resources (Static Device)	9	14	18	23	27	32	36	41	45
Resources (Small Mobile Device)	11	18	23	29	34	40	45	51	56
Resources (Large Mobile Device)	14	21	27	35	41	48	54	62	68

Fig. 2.4 Resource estimation for new cloud service customers (CSCs) with no history, for different requested services

of assigning 0.1 probability value, we have assigned 0.3, which is the average probability of low relinquish. Figure 2.4 shows the unit of resources, and we call it virtual resource value (VRV), being predicted for new customers, for different types of registered services and devices. This unit is then mapped to actual resources (memory, CPU, storage space, bandwidth, etc.), according to the type of service being offered and policies of a particular CSP. For example, a USD 100 cloud storage collaboration service is more I/O intensive. It requires more CPU as well as storage space. The CSP will map VRV 9 to level one of its resource allocation actual mappings. In case the USD 100 service is related to database queries, then only I/O is intensive but not storage, because it requires read-only process. The CSP will perform mappings accordingly. This is how different units of resources are mapped to actual resources, based on the type of service. For USD 100 service, if the accessing device is static, 9 units of resources are reserved. While in case of small mobile device, 11 resources, and for large mobile device, 14 resources are reserved. This is how each type of device is catered according to its requirements. Similarly, for a USD 150 service, if a static device is requesting, 14 units of resources are reserved. On the other hand, small mobile device gets 18 and large mobile device gets 21 units of resources. Fog is then able to handle different types of IoT devices accordingly.

As an example of how mapping can be performed from the above figure, Fig. 2.5 shows the illustrative scenario. CSP maps the resources accordingly on the basis of the type of service and available resource pool. For YouTube service S1, VRV 9

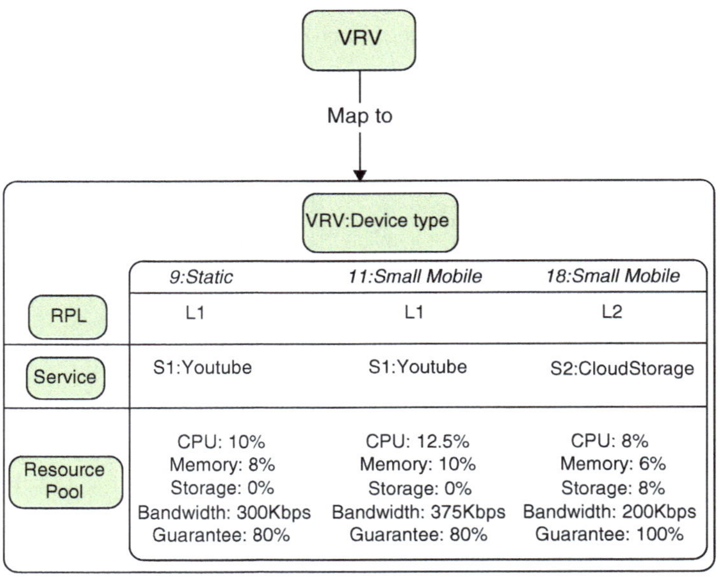

Fig. 2.5 Illustrative scenario of mapping of virtual resource value (VRV) to the resource pool, per the type of service

is mapped to corresponding resource pool level (RPL). Then according to the type of service being provided, the mapping is performed to the actual resource pool. Among the available resources for the service 1, CSP allocates 10% of CPU, 8% of memory, 0% of storage space since storage is not required here, and data rate of 300 Kbps. The guarantee of allocation of these resources is 80%, which means that at least 80% of resources from the mapping are guaranteed. This is only an example. This mapping would vary per the type of service and available resource pool of CSP. For the same service requested by small mobile device, 1.25 times resources are increased for each type of resource.

2.6.2 *Resource Estimation for a Returning Customer*

For a returning/existing customer, fog already has a historical record or RR for each service the CSC has utilized so far. When characteristic of a customer is known, it is more acceptable and reasonable to determine and allocate resources accordingly. In this way, fog will be able to reserve right amount of resources and the chances of losing profit as a result of resource underutilization would be minimized. Figure 2.6 shows five different types of CSCs, having different SR and OR, requesting a particular service *s*. Comparison is shown based on different types of devices, where each CSC requests the same service from different devices. In this example, the result is presented for service price USD 100. The unit is greater for *L* customers, while it is smaller for *H* customers, because of their behavior. Since there are more chances of an *H* customer to relinquish the service(s), hence, more priority and

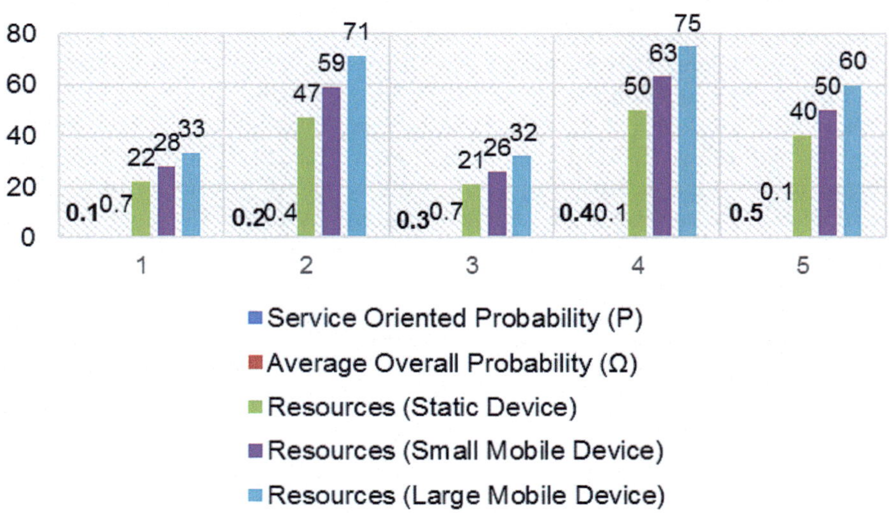

Fig. 2.6 Resource estimation for different types of CSCs, for USD 100 service

quality is provided to the more loyal customer, having *L* probability. In case of CSC 1, having SOP = 0.1 (bold font in the figure) and AOP = 0.7, 22 units of resource are reserved for USD 100 service, when it is requesting from a static device. In case of small mobile device, 28 units of resources would be reserved and for large mobile device, 33 units will be reserved. In case of CSC 2, SOP = 0.2 and AOP = 0.4, 47, 59, and 71 resources are reserved, respectively, for static, small mobile, and large mobile devices. Even though CSC 2 has relatively higher SOP, as compared to CSC 1, but since its AOP is lower than CSC 1, therefore, it gets more resources. CSC 3 has the same AOP (0.7) as that of CSC 1 but has a comparatively higher SOP (0.3). That is why, it gets a smaller amount of allocated resources. CSC 4 and 5 both have an overall perfect loyalty and stability record, having AOP = 0.1. Their resource allocation differs on the basis of SOP. On the basis of currently requested service *S*, CSC 4 is comparatively loyal, having SOP = 0.4, while CSC 5 has SOP = 0.5. This shows that both these types of probabilities have their impact and final decision is made accordingly, which makes it sure that a CSC who has generally been loyal but not so in case of some particular service, or vice versa, gets treated in view of that.

2.7 Conclusion

Fog computing is a middleware entity between underlying IoT nodes and overlying cloud datacenter. Fog is responsible for all initial request-taking and dealing with potential customers. As fog is localized, it is more aware of the underlying nodes, thus, in a more appropriate position to judge how many resources are to be allocated for each kind of CSC. Resource estimation is the most crucial part of overall service provisioning, because correct resource estimation leads to correct resource allocation, which results in correct resource scheduling, and in the end, efficient service provisioning. In this chapter where we extend our previous work on resource estimation modeling at fog layer for IoT nodes, we discuss the key challenges fog has to deal with. One of those challenges is the reliability of CSCs in terms of resource utilization. To keep track of that, we introduce CSC's historical record-based resource estimation. We provide a mathematical model that incorporates customers' give-up probabilities while estimating resources. The algorithm maps the outcome of the historical record ratio module to the type of device that is requesting the resources. Eventually, resources are estimated in accordance with these factors. As a result, dynamic resource estimation is performed which helps in minimizing resource underutilization.

References

1. J. Gubbi, R. Buyya, S. Marusic, M. Palaniswami, Internet of things (IoT): a vision, architectural elements, and future directions. Futur. Gener. Comput. Syst. **29**(7), 1645–1660 (2013)
2. E. Dave, The Internet of Things How the Next Evolution of the Internet Is Changing Everything, Cisco White Paper, April 2011
3. V. Hans, CEO to Shareholders: 50 Billion Connections 2020, Ericsson White Paper, April 2010
4. U. Shaukat, E. Ahmed, Z. Anwar, F. Xia, Cloudlet deployment in local wireless networks: motivation, architectures, applications, and open challenges. J. Netw. Comput. Appl. **62**, 18–40 (2016)
5. M. Aazam, E.N. Huh, Fog computing and smart gateway based communication for cloud of things. In *2014 International Conference on Future Internet of Things and Cloud (FiCloud)*, pp. 464–470, IEEE, August 2014
6. F. Bonomi, R. Milito, J. Zhu, S. Addepalli, Fog computing and its role in the internet of things. In *Proceedings of the First Edition of the MCC Workshop on Mobile Cloud Computing*, pp. 13–16, ACM, August 2012
7. W. Nawaz, K.U. Khan, Y.K. Lee, S. Lee, Intra graph clustering using collaborative similarity measure. Distrib. Parallel Databases **33**(4), 583–603 (2015)
8. M. Aazam, E.N. Huh, Fog computing micro datacenter based dynamic resource estimation and pricing model for IoT. In *29th International Conference on Advanced Information Networking and Applications (AINA)*, pp. 687–694, IEEE, March 2015
9. M. Abu-Elkheir, M. Hayajneh, N.A. Ali, Data management for the internet of things: design primitives and solution. Sensors **13**(11), 15582–15612 (2013)
10. J. Cubo, A. Nieto, E. Pimentel, A cloud-based internet of things platform for ambient assisted living. Sensors **14**(8), 14070–14105 (2014)
11. H. Ning, Z. Wang, Future internet of things architecture: like mankind neural system or social organization framework? IEEE Commun. Lett. **15**(4), 461–463 (2011)
12. S. Chatterjee, S. Misra, Target tracking using sensor-cloud: sensor-target mapping in presence of overlapping coverage. IEEE Commun. Lett. **18**(8), 1435–1438 (2014)
13. C. Sammarco, A. Iera, Improving service management in the internet of things. Sensors **12**(9), 11888–11909 (2014)
14. K. Tei, L. Gurgen, Clout: cloud of things for empowering the citizen clout in smart cities. In *2014 IEEE World Forum on Internet of Things (WF-IoT)*, pp. 369–370, IEEE, March 2014
15. S. Distefano, G. Merlino, A. Puliafito, Enabling the cloud of things. In *Innovative Mobile and Internet Services in Ubiquitous Computing (IMIS)*, 2012 Sixth International Conference on, pp. 858–863, IEEE, July 2012
16. S.J. Stolfo, M.B. Salem, A.D. Keromytis, Fog computing: mitigating insider data theft attacks in the cloud. In *2012 IEEE Symposium on Security and Privacy Workshops (SPW)*, pp.125–128, IEEE, May 2012
17. M. Aazam, E.N. Huh, Dynamic resource provisioning through Fog micro datacenter. In *2015 IEEE International Conference on Pervasive Computing and Communication Workshops (PerCom Workshops)*, pp. 105–110, IEEE, March 2015

Chapter 3
Tackling IoT Ultra Large Scale Systems: Fog Computing in Support of Hierarchical Emergent Behaviors

Damian Roca, Rodolfo Milito, Mario Nemirovsky, and Mateo Valero

3.1 Introduction

The Internet of Things (IoT) represents a phase transition in the evolution of the Internet, characterized by the massive connectivity of endpoint devices (sensors and actuators), and, even more significantly, for the active interaction with the physical world [1]. "Things" capture contextual information from their environment and act upon it. Enabling these functionalities poses new requirements to the underlying infrastructure. Given the highly distributed nature of the "things," they require real-time capabilities such as processing and storage close to where the data is generated. It is widely recognized that security is essential for IoT to reach its full potential.

We are at the dawn of IoT, and envision exciting developments. These developments promise huge benefits although they also offer major challenges. While sensors and actuators occupy today our attention, IoT will evolve organically and get transparently involved in most human activities [2]. This organic evolution of IoT requires the consideration of three dimensions: scale, organization, and contextual awareness [3].

These dimensions are particularly relevant in Ultra Large Scale Systems (ULSS) [4], of which Smart Cities [5] and autonomous vehicles (terrestrial, aerial,

D. Roca (✉) • M. Valero
Barcelona Supercomputing Center (BSC-CNS) and the Universitat Politècnica de Catalunya (UPC), Barcelona, Spain
e-mail: damian.roca@bsc.es; mateo.valero@bsc.es

R. Milito
Senior Technical Leader at Cisco Systems Inc., California, USA
e-mail: rodolfo.milito@gmail.com

M. Nemirovsky
ICREA Senior Research Professor at BSC-CNS, Barcelona, Spain
e-mail: mario.nemirovsky@bsc.es

A.M. Rahmani et al. (eds.), *Fog Computing in the Internet of Things*,
DOI 10.1007/978-3-319-57639-8_3

marine, and submarine) [6] are prime examples. Throughout this chapter we focus on the latter to highlight concepts and architectural foundations. Orchestrating and managing a small set of autonomous vehicles combined with human drivers at modern cities already brings many problems. When that number scales to the majority of circulating vehicles those problems will only aggravate. To ensure a proper operation regime requires handling vast amounts of contextual information for each car to decide its trajectory avoiding collisions.

The stringent latency requirements associated with autonomous vehicles suggests distributed platforms rather than the Cloud for their management [7]. Fog Computing [8] has long recognized the value of extending the Cloud to the edge of the network, bringing networking, compute, and storage resources at different hierarchical levels to respond to the needs of applications and services. Fog addresses the infrastructure and orchestration issues regarding the computational resources [9] (i.e., processing, storage, communications) both at the edge and at different levels of the hierarchy.

A recent paper [10] proposes a fresh approach to design and manage autonomous vehicles (AVs) at scale. More specifically, it proposes Hierarchical Emergent Behaviors (HEB), an architecture that builds on established concepts of emergent behavior and hierarchical decomposition and organization [11]. Useful behaviors can emerge from the application of carefully crafted, well understood, and easy to implement local rules. By leveraging emergent behaviors, HEB brings two major benefits. The first, obvious one, is the bypassing of the need to develop highly complex algorithms. The second, perhaps less obvious, but more important benefit is that HEB's intrinsic flexibility has the ability to handle unanticipated corner cases. The price to pay for these benefits is the need to develop tools (e.g., simulators) to test the emergent behaviors.

The paper on "Emergent Behaviors in IoT" [10] outlined an agenda to deal with ULSS, with emphasis on AVs. This chapter advances the agenda in several significant ways: (a) developing the concept of "emergent behavior primitives," and studying the maneuvers of vehicles exiting a platoon and anticipating to obstacles beyond sensors range; (b) emphasizing the role of Fog Computing as support for HEB communications in general, and facilitating contextual awareness in particular.

The rest of the chapter is organized as follows. Section 3.2 refreshes the main architectural foundations of Fog Computing and its advantages. Section 3.3 revisits and extends HEB. It emphasizes contextual awareness, and introduces the concept of "emergent behavioral primitives [12]," including the synergies between HEB and Fog Computing. Section 3.4 is dedicated to the study of simple, but fundamental primitives: maneuvers by vehicles in a platoon to leave the formation exiting the highway and to anticipate and react to obstacles beyond the sensors range. We discuss the promising results, which suggest that the richness and flexibility of the local rules surpass our expectations. Section 3.5 closes the chapter with the conclusions. Given that this is the first step after the HEB program was announced, we devote some space to discuss open questions and lines of research.

3.2 Fog Computing

Fog Computing is a hierarchically organized architecture of compute, storage, and communication resources that extends from the Cloud to the edge of the network. There are a number of applications and services (e.g., streaming) that can take advantage of Fog. However, from its inception, Fog has been linked to IoT [8, 13].

The true potential of Fog Computing lies in the implementation of a generic multi-tenant platform supporting a wide range of applications simultaneously [14]. Fog breaks down traditional proprietary silos and enables a generic IoT infrastructure, as depicted in Fig. 3.1.

3.2.1 Fog Computing Architecture

A representative architecture of a generic IoT infrastructure is depicted in Fig. 3.2. At the lower levels are the "things," sensors and actuators responsible of gathering information, and acting on the environment. The next layer is formed by heterogeneous Fog nodes, which constitute the aggregation points. The "things" and the nodes communicate mostly through wireless technologies, since both "things" and nodes can move. Due to the Fog nodes' wide geographic deployment and their location, they can offer resources in real time by processing the data close to where it is generated. The Fog nodes form an interconnected hierarchy. In most cases higher nodes have larger pools of resources at the cost of an increased latency. The Cloud constitutes the highest layer, offering a large pool of resources at low-cost without any latency guarantees.

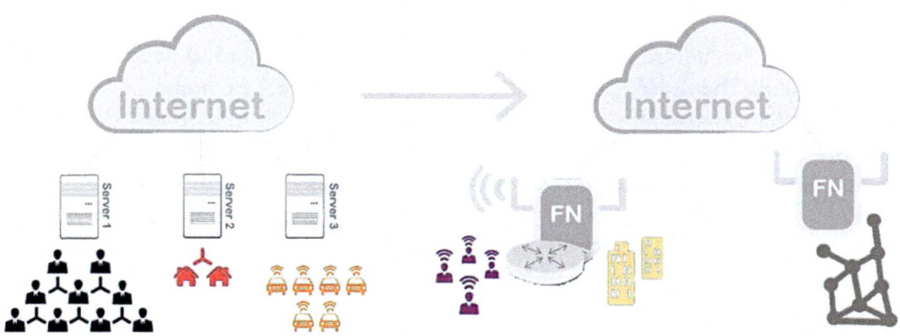

Fig. 3.1 Moving from a silo-based implementation to a generic Fog infrastructure capable of supporting multiple applications simultaneously

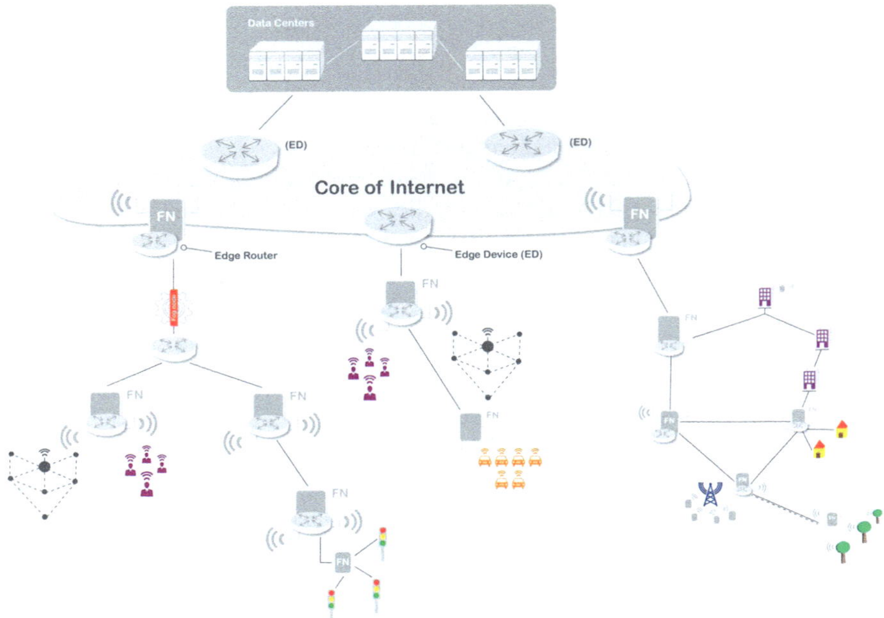

Fig. 3.2 Illustrative example of a generic Fog-based infrastructure serving multiple IoT applications. Fog nodes are interconnected forming a hierarchy

3.2.2 Fog's Role Within ULSS

Having laid out the foundations of the architecture we can explain the role that Fog nodes can play within ULSS. These systems can exploit the locations of the Fog nodes and their hierarchical organization to communicate and to become aware of their context. Thanks to their interconnected architecture, Fog nodes have visibility over a wider geographical range than the one available to individual "things." For example, in the AVs use case, a Fog node at the bottom of the hierarchy aggregating data from a set of vehicles has information of a broader area than each individual vehicle. In contrast, each car manages contextual information on a narrower area limited by the range of its sensors and its immediate neighbors in the road. In addition to the broader geographical range, Fog nodes can provide a deeper vision in time. For example, historical information on the traffic conditions.

This visibility places the nodes as notorious information distribution points, fact reinforced by its hierarchical organization (the higher the node, the higher its scope). Then, Fog nodes can transmit information such as the road conditions to optimize cars' trajectories in real time.

Following the reverse process, "things" can use the Fog nodes for their own functionalities. "Things" under a Fog node coverage can use the computational resources of that node to analyze measurements or perform other tasks. When "things" move out of the range they get disconnected from that node. If another node is available in the next location, the same process can continue. This technique eliminates the need of migrating data from one node to the next one because the "thing" itself carries the necessary information.

3.3 Hierarchical Emergent Behaviors, a Fresh Approach for ULSS

ULSS present architects and developers unique challenges not only because of their massive scale (the number game [4]) but also because of their richness (the diversity of the possible scenarios). The first paper laid HEB's architectural foundations and its organizational principles. This chapter offers concrete evidence of HEB's value by exploring in-depth basic maneuvers of autonomous vehicles.

3.3.1 HEB Architecture

HEB builds up on top of two concepts, namely emergent behaviors and hierarchical decomposition. The former concept induces behaviors through a set of local rules that define interactions between neighboring "things." The latter organizes complex systems into different levels, each abstracting the essential functionalities of the previous one while maintaining its functionalities. The combination of both areas results in "things" applying lightweight rules that define their interactions with other "things" and also with their environment, as depicted in Fig. 3.3.

Within each rule there is a set of hyper-parameters, e.g., separation distance. Their values can be established at configuration time or dynamically. The main outcome of this technique is the leveraging of the decision process to the "things" themselves. Rather than attempting to explicitly program all potential scenarios in advance (daunting task in the huge state space), the system has the capacity to react effectively to unanticipated situations.

HEB's potential best manifests itself when the principal actors in the system are mobile (terrestrial, aerial, marine, and submarine vehicles). Within this class of applications we focus on terrestrial AVs. Rather than explicitly programming the vehicles, we develop simple, well understood local rules that regulate the interactions between neighboring vehicles as well as with the external world.

A set of local rules does not get automatically tagged to a desired behavior. A careful selection process is required. An expressive, realistic simulation platform, coupled with carefully designed experiments, is a promising tool to experiment with rules, their induced behaviors, and the system reaction to unanticipated events.

This new paradigm demands intra and inter-level communications. Rules depend on those interactions to define the "things" behaviors. Without proper commu-

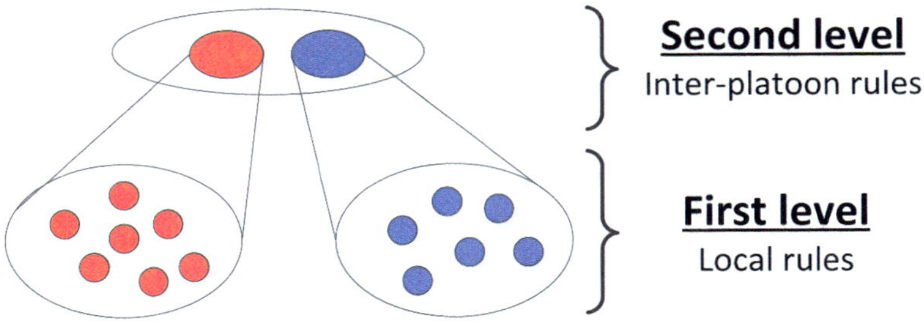

Fig. 3.3 Representation of the HEB paradigm with two levels. The first level rules apply to level 1 elements (e.g., AVs) inducing a platoon behavior. The second level applies inter-level rules over the previous level behaviors (e.g., platoons) to enable complex functionalities

nications between "things" and the proper set of sensing capabilities of their environment, behaviors cannot emerge. In the AVs application, these communication capabilities include the ability to exchange information with neighboring cars and roadside units (RSU) and to measure their relative positions and velocities (LIDAR, cameras). In addition, each vehicle is aware of its own location and velocity (GPS, accelerometers). The sensors are already available, the communication protocols (e.g., DSRC [15]) well developed and tested, and given the current state of excitement in the field, the infrastructure will be deployed in the not too distant future.

3.3.2 HEB, the Next Phase

The application of the three original rules from Reynolds [16] to a set of autonomous vehicles results in the formation of a platoon [17] without explicitly program that behavior. However, these rules do not specify the absolute velocity of the group. Platoon absolute velocity is defined as the absolute average velocity of all the vehicles forming the platoon. This velocity is a crucial metric in autonomous vehicles and highly depends on the context, including the quality of the road, weather conditions, vehicle density, maneuvers, and neighboring platoons among others.

The above considerations strongly suggest the need to define the policy not only in terms of local rules. At the end, a policy maps the information state of the system into an admissible set of decisions. For AV HEB, a policy at any given level of the hierarchy includes:

- Local rules pertaining to the hierarchical level.
- The set of hyper-parameters associated with those rules. This set not only includes parameters such as the rules' weights and separation distance but also velocity applied to each level (i.e., first level refers to average speed of the cars, at the second level is a vector of velocities for each platoon).

- Contextual information. The challenge is to capture in a succinct way the critical information. This requires analysis and careful experimentation. The issue is the required degree of granularity. Contextual awareness includes car density, weather conditions, road conditions, platoon regime, etc.

Architects can define a policy portfolio with well-known emergent behaviors to implement. Given that contextual information is captured in the policies, the selection process becomes a simple, even a trivial one. There are only a few admissible policies for a given informational scenario. Then, the first set of policies to define are the so-called emergent behaviors primitives.

3.3.2.1 HEB Primitives

By primitives we understand basic operations required by vehicles within a platoon. Right now we focus on first level behaviors, but the same concept applies to any level within HEB. Vehicle maneuver without collision or handling autonomous cars that want to take an exit in a highway constitute primary examples of a primitive [17]. Simple as they sound, this requires consideration of different aspects and interactions of HEB components:

- Communications between different entities: (1) vehicle to vehicle, (2) vehicle to RSU, and (3) distribution of functionalities within the platoon
- Vehicle announcement of its intent
- Non-intersecting exit trajectories whether one or multiple vehicles leave the platoon
- Emergent behaviors at play: current operating rules, their hyper-parameters, and new individual behavior (i.e., exiting the platoon) affecting the emergent behavior.

Taking a closer look to these aspects of the maneuver without collision, we observe the effect of the emergent behaviors through the separation rule between "things" and obstacles. In this case, a single rule provides us the primitive objective if the proper sensing capabilities are satisfied in each moving vehicle.

The exiting highway maneuver requires more considerations. Although each vehicle does not know the number of vehicles in the platoon nor has membership awareness, it may notify to its neighboring cars its exit. There is a fundamental reason for this notification: to avoid an undesired behavior with the entire platoon unconsciously following the exiting vehicle/s.

Communications constitute a key element to ensure a satisfactory maneuver. Efficient exiting strategies necessarily rely on contextual information. It is useful to distinguish between permanent information (coordinates of the exit, proximity to other exits, etc.) from real-time information (state of the road, congestion level at the exit, weather conditions, speed of the platoon, vehicular density).

Fog Computing is of great help, from the compute and storage capabilities of the RSU at the edge, to the exchange of real-time information along the RSUs.

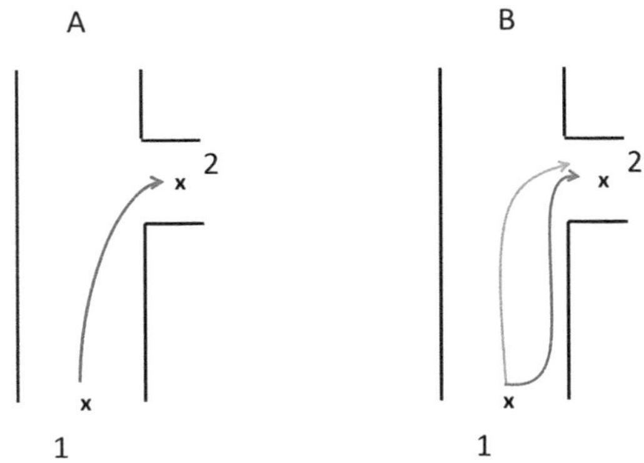

Fig. 3.4 *Left figure* shows a specific trajectory that a vehicle must closely follow. Instead, the *right figure* shows the freedom HEB leverages to the vehicles to decide their trajectory. Vehicles are capable of taking best decisions since they have all the contextual information

Fog nodes then become the RSUs in the roads providing their capabilities to the vehicles while building applications on top of their contextual information (i.e., smart guidance systems). Another alternative could be to use the same vehicles to detect and classify the lanes [18].

Last but not least, the use of non-intersecting trajectories in the 3-dimensional space is a must. In traditional solutions with explicitly programmed behaviors, a central orchestrator determines each vehicle's trajectory and makes sure no collision happens. Instead, HEB defines a rule to prevent collisions and gives freedom of choice to the vehicles to decide the best trajectory based on their contextual information. Figure 3.4 depicts these differences between the two methodologies. In opposition to preset trajectories, HEB approach creates local rules that lead to behaviors emerging in the form of trajectories.

3.4 Two Autonomous Vehicles Primitives Case Study

AVs constitute one of the clearest HEB applications. The mobility of the vehicles, the constantly changing situations (i.e., road conditions, weather, traffic conditions), and the number of cars provide a rich set of interactions from which behaviors emerge.

The universe of AV maneuvers is the composition of a vast set of "primitives." We approach the validation of the HEB architecture by creating and examining in detail a rich "library" of primitives. Each primitive is defined by a goal, and it is self-contained in that it has the ability to reach said goal. We envision the creation of complex scenarios by concatenating primitives. More precisely, we will consider richer goals that are the composition of simpler goals, and achieve them by concatenating primitives. This section, which is the first step in this direction, focuses on two primitives that rely on Fog nodes deployed as RSUs.

While a primitive is defined by a goal, its full characterization necessitates the specification of the rules that facilitate the achievement of the goal.

3.4.1 *Methodology*

We chose the Processing simulator [19] to perform our analysis of emergent autonomous vehicles. We take as base the set of rules and the environment from the original HEB article where each vehicle determines its trajectory based on four rules that guide its local interactions (R_1 Alignment, R_2 Separation, R_3 Cohesion, and R_4 Destination).

In this first validation effort we limit our attention to first level primitives, that is, primitives that relate to the formation and lifetime of a platoon, and disable the second level rules, which pertain to the interactions between platoons.

Triangles represent the vehicles. Obstacles define the shape of the highway with a sideway where the primitives are tested, shown as black dots in the canvas.

3.4.2 *Emergent Autonomous Vehicle Primitives*

The two primitives under evaluation are the exiting maneuver in a highway and anticipating and reacting to obstacles beyond the sensors range. There are many ways of implementing a primitive. In the following sections we perform a preliminary analysis and present tentative solutions that satisfy the objectives of these two primitives.

3.4.2.1 Exiting Maneuver

In this primitive one or more vehicles in a platoon decide to leave the platoon and exit the highway. We consider the primitive is accomplished satisfactorily if the vehicles exit without collisions or hazardous maneuvers and the rest of the platoon continues the journey unperturbed [20]. Exiting brings forth the interplay between emergent behavior, classical trajectory design methodologies, and inter/intra-layer communications:

- Communications: they were explained before in detail, mainly focused on V2V and V2I [21] enhanced with each vehicle's sensing capabilities.
- Emergent behaviors: Platoon behavior (induced through the three original Reynolds rules) and moving objects start roughly at the same speed along the road. As the exiting maneuver proceeds, the velocity vector of the moving object changes in direction and magnitude, but not in a brusque way.
- Trajectory design: Vehicles in the platoon as well as exiting vehicles are interacting but autonomous decision makers, in that they sense the environment and react accordingly. The strategy of fixing exiting trajectories and relying on the collision avoidance ability of vehicles in the platoon seems sound and straightforward. This strategy regards only the platoon vehicles as interacting autonomous decision makers, as depicted in Fig. 3.4.

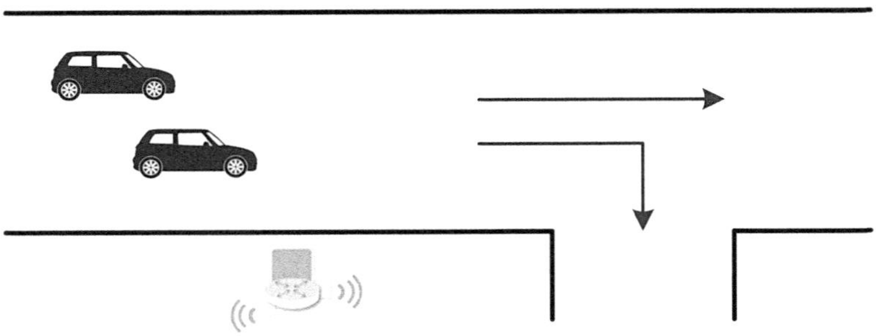

Fig. 3.5 Scenario to evaluate the exiting maneuver. It consists of a highway with an exit. The autonomous vehicles can either exit or continue in the highway based on their final destination. The RSU deployed as a Fog node assists with the contextual information

Among the possible implementations of the exiting primitive we analyze three possibilities of different complexity and observe their impact on the behaviors of interest. The first implementation starts with vehicles notifying their intent to leave the platoon. Since there is no central control of the platoon or membership awareness each vehicle needs to handle the notification to its surrounding vehicles. An intuitive way of notifying its intention is to change the vehicle's role within the platoon. Instead of being perceived as a vehicle, and therefore subjected to the three platoon rules (R_1, R_2, and R_3), perception changes to that of a moving obstacle. In this case, the rest of the vehicles within the platoon avoid it by simply following the non-collision rule (R_2). This technique results in exiting vehicles creating a virtual path within the platoon till they make their exit. The vehicle's new role allows it to leave the platoon and take the desired exit without compromising the platoon behavior for the remaining vehicles.

The challenge is to effect that change of role (from a vehicle in the platoon to moving object) without affecting the local rules or resorting to a central orchestrator. Visualize the scenario in which a platoon of vehicles moves along a highway as depicted in Fig. 3.5. A RSU notifies the platoon of the existence of an exit ahead. The RSU is actually a Fog node, part of a full Fog hierarchy deployed along the highway. The Fog node keeps contextual information including obstacles in the road ahead, congestion levels, weather conditions in the area, etc., as part of the rich information exchanged with other Fog nodes, both at the same and higher hierarchical levels. Hence, the Fog can extend a vehicle "vision" beyond the capabilities of the on-board sensors.

A vehicle decides to take the forthcoming exit, and broadcasts its neighbors the change of its role, from a peer in the platoon, to mobile obstacle. It does so through the V2V communication channel (e.g., DSRC). From that moment on, that vehicle is perceived as an obstacle by any vehicle happens to be in its neighborhood. As the exiting vehicle maneuvers, its neighborhood changes, but as the notification of its role keeps active, the new neighbors keep away from it. Hence, the exiting vehicle carves a wormhole through the platoon that leads to the exit.

The same methodology applies when more than one vehicle wants to take the exit. In this case, each exiting vehicle acts individually and it is not coordinated with the other exiting entities. We exploit the power of the rules and their flexibility. Since each vehicle avoids obstacles (R_2), there will be no collision among vehicles whether they are part of the platoon or they are leaving. There is no need to implement costly orchestration mechanisms to anticipate all possible situations in micro detail, we just give basic rules and let the vehicles decide what is best for them. The result is a set of vehicles "leaving" the platoon and taking the exit, while the rest of the platoon moves along the highway to its destination.

The second implementation uses on a more direct approach based on the rules and their hyper-parameters. This solution does not require extra communications (i.e., notifications) to achieve the primitive's objective. When the RSU announces the exit that one or more vehicles want to take, the exiting vehicles modify their destination target in R_4 and simultaneously modify the weight associated with that rule. Recall that weights define priorities among the rules that determine the local behavior. The separation rule (R_2) still keeps the highest priority to ensure no collision happens but the destination rule dominates (R_4) over the rest (R_1 and R_3).

While this approach also produces the desired result (vehicles exiting the highway without collisions), we observe differences in the behaviors, which may affect the time it takes to exit the highway. Giving priority to the destination rule over the traditional platoon rules ensures that the vehicles take the exit instead of continuing as part of the platoon. Similar groupings based on destination were analyzed by Hall et al. [22]. What changes with respect to the previous case is how the local rules apply. While the previous technique is based on vehicle to object interactions, the second one relies on rules between vehicles.

We experimented with a third approach, which is in fact a particular case of the previous one: exiting vehicles modify their destination targets, but they do not alter the weight of the rule. The fact that this approach meets the goals of the exiting primitive highlights the surprising expressiveness of the local rules. Adding a target destination rule (R_4) we can induce many useful behaviors. Besides the obvious behavior of reaching a destination, vehicles with different targets can form a common platoon and later one split to reach both destinations.

Figure 3.6 depicts a temporal representation of the exiting maneuver implementing the third technique (the results from the previous two are the same except the aforementioned differences). The left figure shows the platoon at the beginning of the highway just before crossing the RSU that communicates the forthcoming exit. The center figure shows some vehicles "leaving" the platoon. Finally, the right figure shows the small platoon of exiting vehicles as well as the platoon of remaining vehicles in the highway. Simulation details not in the figure show exiting cars moving to the edge of the platoon, positioning themselves for a smooth exit, without vehicles crossing their paths. This behavior, not explicitly programmed, emerged naturally from the local rules. It is actually the result of some rules dominating others (in this case R_4).

The policy emerging with the third technique has considerable degrees of freedom. Consider, for instance, the rare case in which an exiting vehicle finds

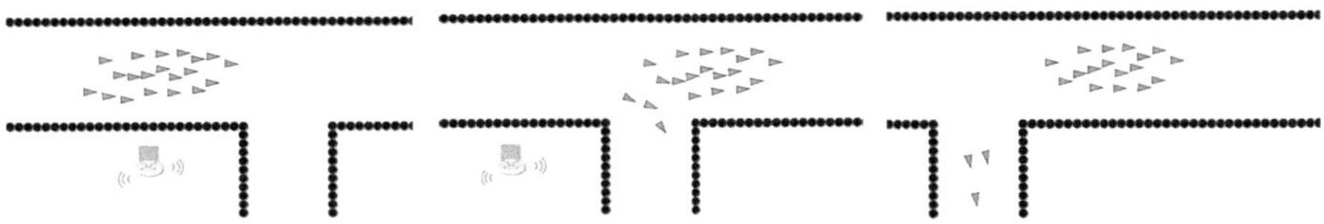

Fig. 3.6 Sequential representation (from *left* to *right*) of a platoon executing the exiting primitive assisted by the RSU

the exit suddenly blocked by vehicles ahead. The vehicle cannot force its way out because it is not acting as a mobile obstacle, but rather as a peer of the other vehicles. The car can head back, and rejoin the platoon. Note that this would not be the case with the previous techniques, because they are more aggressive.

Another remarkable case happens if the platoon does not have a specified target destination at the end side of the highway. In this scenario, when the exiting vehicle executes its maneuver, the rest of the platoon can follow it. This behavior is not problematic though because each vehicle always has a target destination.

A simple set of four rules provides a wide range of useful behaviors that are very robust. In addition these lightweight rules demonstrate an incredible flexibility and simplicity. Fog nodes contribute to handle the contextual information. In this case it only transmits the target destination that will make the vehicles take the proper sideway. We have seen three different ways of implementing this primitive but there are more options that can be part of the policy portfolio (rules, hyper-parameters, contextual information) and can be reused for other applications, reducing the deployment time.

3.4.2.2 Anticipating and Reacting to Obstacles Beyond the Sensors Range

In this primitive a vehicle or a set of them is circulating in a highway and there is an obstacle beyond the on-board sensors range. The main objective is to anticipate its detection and react accordingly to overcome it without compromising the safety of the driving. Overcoming obstacles brings forth the same areas as the exiting maneuver with the difference that this primitive directly targets the hyper-parameters within the emergent behavior rules.

This scenario is slightly different from the previous ones. We have a platoon moving along a highway as depicted in Fig. 3.7. Along the road there are a set of RSUs that gather information about the road conditions, weather, and traffic among others. In summary, these Fog nodes manage contextual information related to the highway. These nodes are organized hierarchically to capture the information of a wider area. Figure 3.7 also depicts the virtual architecture they conform with two different levels. The first level is formed by the RSUs closer to the highway, the nodes that physically deal with the vehicles. On the second level there is a single RSU that aggregates the information from the previous level. This level 2 node has a wider scope but its granularity is coarser.

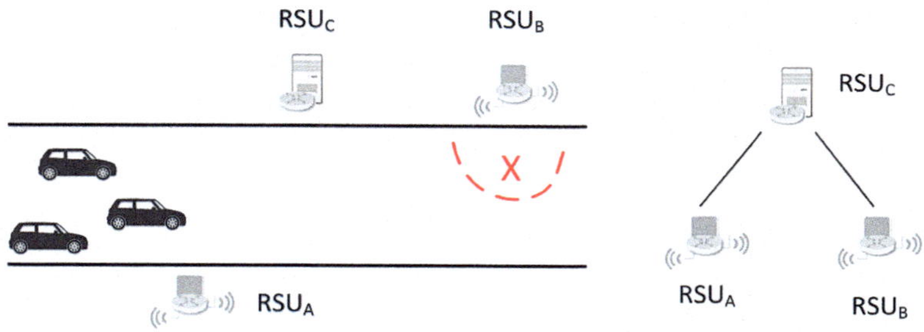

Fig. 3.7 Scenario to anticipate and react to obstacles beyond the sensors range. It consists of a highway with a three RSUs organized hierarchically. The *right side of the figure* details this architecture with two nodes at the bottom that directly communicate with the "things," and a higher node to aggregate all the data. The *cross* represents a temporal blockade in the road

This hierarchical RSU configuration provides information to the vehicles that traverse the boundaries of locality. While vehicles only sense its closest surroundings, RSUs have a larger scope and can transmit that information to the vehicles. In this situation, vehicles can prepare for the upcoming obstacle or other events. The procedure is as follows: RSU$_B$ senses the blockade and besides notifying it to the vehicles within its range, it sends this information to the higher layer node, RSU$_C$. Then, this node can transmit the information to the other lower level RSUs to take proper actions. In this scenario, RSU$_C$ sends the notification to RSU$_A$ that is the node at the left. Now, RSU$_A$ has the information on the road status ahead and can notify it to the nearby platoon.

To optimize the reaction of the platoon to the blockade, RSU$_A$ acts upon the hyper-parameters modifying the separation distance between the vehicles and also their speed. This action influences the emergent behavior in real time. We need to be careful not to augment this distance over the sensing capabilities of each vehicle, fact that will preclude the formation of the platoon. On the other hand, too small a value could result in collisions. Other factors such as the number of lanes in the road also place constraints over this hyper-parameter. In this analysis we keep this distance between acceptable boundaries that do not compromise the behavior. Reducing the distance we induce a compact platoon that can overcome the obstacle easily. To induce proper trajectories, RSU$_A$ establishes a destination point through R_4 to overcome the blockade smoothly. We are influencing the behaviors through its rules with the final objective of reducing the reaction time.

Figure 3.8 presents a temporal sequence on how the cars execute this primitive. The left figure presents the initial position of the platoon moving the highway in normal operation regime mode. The center figure shows the modified behavior after RSU$_A$ has modified the separation hyper-parameter, the speed, and the destination target of the platoon. As you can observe, the platoon now is more compact and the vehicles move closer between them. The right figure shows the platoon overcoming the blockade previously notified to it. Finally, once the platoon totally surpassed the

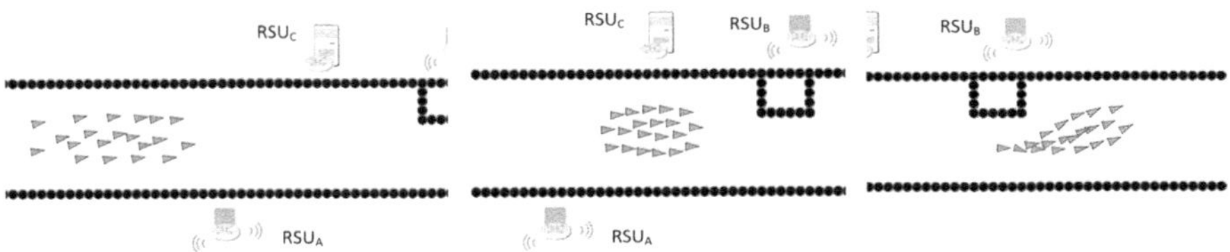

Fig. 3.8 Sequential representation (from *left* to *right*) of a platoon executing an anticipated reaction to obstacles beyond the sensors range

blockade RSU_B reestablishes the original separation distance, speed, and triggers the original target destination.

The advantages of such a primitive include a reduction of the time to overcome situations such as partial road blockades and to show how behaviors can be influenced by the contextual information. In this specific situation we achieve it through a hierarchy of RSUs although there are other solutions. The reaction time reduction comes from how vehicles face the blockade beyond their sensors range. If they are not prepared, a part of the platoon uses the blocked lanes in front of them. Once they sense that obstacle they will change their trajectory to avoid it, but they may have to wait till the vehicles on the clear lanes pass through it. The other possibility is even slower, when both vehicles intersect and the absolute velocity drastically diminishes. Instead, if the platoon is more compact and there are fewer cars on the blocked lanes the reaction time is smaller.

3.4.2.3 Concatenation of Primitives to Express Complex Behaviors

Judiciously chosen local rules, though simple to understand and implement, have the amazing capability of inducing behaviors not explicitly enunciated. Local rules are also flexible and expressive, enabling the creation of primitives through minor tweaking and additions, as shown in the previous section. We observe that the exiting maneuver and obstacle anticipation primitives presented are built on top of a proto-primitive, the platoon formation. This observation suggests that complex behaviors can be achieved by chaining primitives.

Here is an outline of the envisioned methodology, to be developed in detail:

- Create a library of primitives.
- Concatenate primitives to design complex behaviors.
- Test the results through extensive simulations using realistic simulation tools.

The fairly extensive literature on self-driving vehicles that follow prescriptive designs [23] methodology can be leveraged in two ways: (a) it suggests a list of primitives and complex behaviors to consider; (b) the use cases can be used as baselines to compare with the HEB methodology.

3.5 Conclusions

In this chapter we validated the HEB concepts introduced in [10] with two specific use cases in the AVs domain. We advanced further the HEB approach by introducing the concept of "primitives" and their concatenation.

The use cases discussed highlight the role of Fog Computing in supporting the vehicle to RSU communication, as well and providing essential contextual information that extends beyond the immediate neighborhood. This chapter contributions illuminate the fact that HEB remains in a nascent stage, and that rich lines of research remain open:

- The primitive led design concept reinforces the need of work on rich, realistic simulators, able to incorporate real world scenarios and data.
- The concatenation of primitives deserves attention, both at a rigorous development of the methodology and the practical building of a portfolio of complex behaviors.
- Machine learning promises to play a major role in the tuning of the hyper-parameters at the core of the local rules.
- The domain of autonomous vehicles extends well beyond terrestrial vehicle. Aerial, marine, and submarine autonomous vehicles own idiosyncrasies are worth exploring with the HEB approach.

Acknowledgements Damian Roca work was supported by a Doctoral Scholarship provided by Fundación La Caixa. This work has been supported by the Spanish Government (Severo Ochoa grants SEV2015-0493) and by the Spanish Ministry of Science and Innovation (contracts TIN2015-65316-P).

References

1. L. Atzori, A. Iera, G. Morabito, The internet of things: a survey. Comput. Netw. **54**(15), 2787 (2010)
2. R. Milito, short video in the September 2016 issue of computing now. [Online]. Available: https://www.computer.org/web/computingnow/archive/iot-data-and-context-discovery-september-2016
3. G.D. Abowd, A.K. Dey, P.J. Brown, N. Davies, M. Smith, P. Steggles, Towards a better understanding of context and context-awareness, in *International Symposium on Handheld and Ubiquitous Computing* (Springer, Berlin, 1999), pp. 304–307
4. L. Northrop, P. Feiler, R.P. Gabriel, J. Goodenough, R. Linger, T. Longstaff, R. Kazman, M. Klein, D. Schmidt, K. Sullivan et al., Ultra-large-scale systems: the software challenge of the future, DTIC Document, Technical report, 2006
5. G.P. Hancke, G.P. Hancke Jr et al., The role of advanced sensing in smart cities. Sensors **13**(1), 393–425 (2012)
6. M. Gerla, E.-K. Lee, G. Pau, U. Lee, Internet of vehicles: from intelligent grid to autonomous cars and vehicular clouds, in *2014 IEEE World Forum on Internet of Things* (IEEE, Piscataway, NJ, 2014)

7. K. Sasaki, N. Suzuki, S. Makido, A. Nakao, Vehicle control system coordinated between cloud and mobile edge computing, in *2016 55th Annual Conference of the Society of Instrument and Control Engineers of Japan (SICE)* (IEEE, Piscataway, NJ, 2016), pp. 1122–1127
8. F. Bonomi, R. Milito, J. Zhu, S. Addepalli, Fog computing and its role in the internet of things, in *Proceedings of the First Edition of the MCC Workshop on Mobile Cloud Computing* (ACM, New York, 2012), pp. 13–16
9. S. Shin, S. Seo, S. Eom, J. Jung, K.-H. Lee, A pub/sub-based Fog computing architecture for internet-of-vehicles, in *2016 IEEE International Conference on Cloud Computing Technology and Science (CloudCom)* (IEEE, Piscataway, NJ, 2016), pp. 90–93
10. D. Roca, D. Nemirovsky, M. Nemirovsky, R. Milito, M. Valero, Emergent behaviors in the internet of things: the ultimate ultra-large-scale system. IEEE Micro **36**(6), 36–44 (2016)
11. H.A. Simon, *The Architecture of Complexity* (Springer, New York, 1991)
12. M.J. Mataric, Designing emergent behaviors: from local interactions to collective intelligence, in *Proceedings of the Second International Conference on Simulation of Adaptive Behavior* (1993), pp. 432–441
13. M. Yannuzzi, R. Milito, R. Serral-Gracià, D. Montero, M. Nemirovsky, Key ingredients in an IoT recipe: Fog computing, cloud computing, and more Fog computing, in *IEEE 19th International Workshop on CAMAD* (IEEE, Piscataway, NJ, 2014)
14. Open Fog Consortium. [Online]. Available: https://www.openfogconsortium.org/resources/
15. J.B. Kenney, Dedicated short-range communications (dsrc) standards in the united states, *Proc. IEEE* **99**(7), 1162–1182 (2011)
16. C.W. Reynolds, Flocks, herds and schools: a distributed behavioral model, in *ACM SIGGRAPH Computer Graphics* (ACM, New York, 1987)
17. P. Varaiya, Smart cars on smart roads: problems of control. *IEEE Trans. Autom. Control* **38**(2), 195–207 (1993)
18. J. Melo, A. Naftel, A. Bernardino, J. Santos-Victor, Detection and classification of highway lanes using vehicle motion trajectories, *IEEE Trans. Intell. Transp. Syst.* **7**(2), 188–200 (2006)
19. Processing simulation framework. [Online]. Available: https://processing.org/
20. A. Hsu, F. Eskafi, S. Sachs, P. Varaiya, Design of platoon maneuver protocols for IVHS, in *California Partners for Advanced Transit and Highways (PATH)* (University of California, Berkeley, 1991)
21. Its vehicle to infrastructure resources. [Online]. Available: http://www.its.dot.gov/v2i/
22. R. Hall, C. Chin, Vehicle sorting for platoon formation: impacts on highway entry and throughput. Transp. Res. Part C: Emerg. Technol. **13**(5), 405–420 (2005)
23. E. Frazzoli, M.A. Dahleh, E. Feron, Real-time motion planning for agile autonomous vehicles. J. Guid. Control Dyn. **25**(1), 116–129 (2002)

Part III
Services of the Fog Layer

Chapter 4
The Present and Future of Privacy-Preserving Computation in Fog Computing

Patrícia R. Sousa, Luís Antunes, and Rolando Martins

4.1 Introduction

The increasing attacks on users' privacy reveal the economical importance of sensitive data for both the companies and criminals. The need to reduce time-to-market has led companies to deploy edge computing systems without security and privacy by design. The exponential growth of the data generated by this type of systems is chronically exacerbating the problem.

Fog computing is one of the most predominant examples of edge systems, and is one facet of the overarching concept that is the Internet of Things (IoT). It can be best described as the relocation of computing, preferably closer to devices near the end user. Current security and privacy approaches used in cloud computing infrastructures are not appropriate for the underlying requirements of fog computing, namely their intrinsic decentralized nature. Furthermore, privacy-preserving frameworks are still not available in public cloud computing, and are an active field of research. Current approaches use pseudo-anonymization techniques that can be de-anonymized with the aggregation of multiple sources of information.

Decentralized in nature, fog computing offers a pathway that we argue can be used to create novel privacy-preserving techniques. By keeping the data near the end user it provides a natural barrier to large scale data collection performed by big-data/analytics based companies. By itself, edge systems do not provide the necessary mechanisms to meet this goal. To fill this gap, there are some privacy primitives that we want to explore and potentially combine.

The first relies on secure multi-party computing (SMPC), that can be used to compute responses based on confidential data, so that when the protocol is completed users know only their own input and the answer. Concurrently, blockchains

P.R. Sousa (✉) • L. Antunes • R. Martins
Department of Computer Science, University of Porto, Rua do Campo Alegre, Porto, Portugal
e-mail: psousa@dcc.fc.up.pt; lfa@dcc.fc.up.pt; rmartins@dcc.fc.up.pt

A.M. Rahmani et al. (eds.), *Fog Computing in the Internet of Things*,
DOI 10.1007/978-3-319-57639-8_4

offer public ledgers that reduce and potentially eliminate the presence of centralized trusted entities. As an example, within the Bitcoin virtual currency, a blockchain is the data structure that represents a financial accounting entry or a record of a transaction. Each transaction is digitally signed with the purpose of guaranteeing its authenticity and ensuring that no one adulterated it, so that the record itself and the transactions within it are considered of high integrity.

In this chapter, we focus on these two different approaches as a means to create solutions to enhance privacy-preserving approaches for IoT. We also explore the combination of blockchain and multi-party computation techniques on the edge.

The next sections of this chapter are organized as follows: Sect. 4.2 presents the blockchain and their limitations in the integration with IoT. Also, the same section presents the IoT, fog computing, and blockchain concepts together with their respective applications. In Sect. 4.3, we describe the multi-party computations, the description and comparison between MPC frameworks, and also some future research directions of this field. Section 4.4 presents both multi-party computation and blockchain and the applications of the concepts together. Also, we describe the future research directions of the combination of the two technologies. Lastly, we have a summary of this chapter in Sect. 4.5.

4.2 Blockchain

Blockchain is a decentralized ledger of all Bitcoin[1] transactions across a peer-to-peer network, growing as miners add new blocks to it in order to record new transactions. The nodes within the network validate the transaction and the user's status by using known algorithms to ensure that the same Bitcoins were not spent previously, thus eliminating the double expense problem. A verified transaction can involve cryptocurrency, contracts, records, or other information. Once the transaction is verified it gets combined with other transactions in order to create a new data block within the public ledge. The new block is then added to the existing blockchain, in a way that is permanent and immutable. Only at this point is the transaction considered completed [2].

Using this technology, users can confirm transactions without the need for a central certifying authority, normally enforced by central banks. Other possible applications include fund transfers, settling trades, and voting.

[1] Bitcoin is a digital currency and online payment system, also called digital cash. It works in a decentralized way, that uses peer-to-peer to enable payments between parties without the need of mutual trust. The payments are made in Bitcoins that are digital coins issued and transferred by the Bitcoin network [1].

4.2.1 *Blockchain Limitations for IoT*

The integration of blockchain in IoT is not straightforward, since the majority of IoT devices are resource restricted and low latency is desirable. Also, there are a large number of nodes in IoT networks and devices are bandwidth-limited. In order to successfully integrate blockchain with IoT several critical challenges will have to be addressed, namely:

- Mining is computationally intensive;
- Mining of blocks is time consuming;
- Blockchain scales poorly as the number of nodes in the network increases;
- The underlying BC protocols create significant overhead traffic [3].

4.2.2 *The Internet of Things, Fog Computing, and Blockchain*

Despite the fast paced development of new architecture frameworks for IoT, privacy and security remains a second class citizen, leading to several open privacy and security challenges. There are several advantages of using blockchain technology that, in theory, can potentially help to improve privacy and security, majorly through decentralization. As described in [3], user identities must be kept private and this can be accomplished when using a blockchain. As they are decentralized by design, blockchains offer scalability and robustness by using the resources from all participating nodes, and in the case of IoT, from all devices. In the process it also eliminates many-to-one traffic flows, which reduces delays, and overcomes the problems associated with the presence of single point-of-failure [4].

However, there are problems that need to be solved in order to allow practical usage of blockchain. For example, IoT devices normally have limited resources that are not enough to properly support cryptocurrency mining due to its computational cost. It is desirable for IoT applications to have low latency and low traffic in the network. Mining of blocks is time consuming and creates significant overhead traffic, which is undesirable. Moreover, blockchain does not properly scale with the ever-increasing introduction of nodes in the network [3].

In order to solve the current limitations within IoT that would provably compromise a seamless integration with blockchains, a new paradigm must be used. We argue that fog computing is a prime candidate be employed in this scenario. One of its main goals is to deal with the current limitations of public cloud computing when dealing with services and applications that requires very low latency, "local awareness," and mobility (including vehicular mobility).

Follows the presentation of approaches that explore the integration of between IoT, blockchain and fog computing. We will describe their applications and how they address the aforementioned challenges.

4.2.3 IOTA

IOTA is designed to be as lightweight as possible, unlike the complex and heavy duty blockchain operations of Bitcoin. IOTA is a novel transactional settlement and data transfer layer for IoT. The first part in "IOTA" emphases the importance that is conceded to the IoT. It is based on a new distributed ledger, the Tangle, which overcomes the inefficiencies of current blockchain designs and introduces a new way of reaching consensus in a decentralized peer-to-peer system [5].

With the growth of the number of devices that exist in the IoT, that can reach tens of billions of connected devices in the next decade, one of the main needs are the interoperability and resource sharing. For this, IOTA enables companies to explore new business-to-business (B2B) models by making every technological resource a potential service to be traded on an open market in real time and without fees.

Recently, new approaches for IoT have been proposed with the introduction of Fog and Mist[2] [7]. The main goal of these new paradigms is to decrease the network latency to cloud servers that can located far away from end-devices. Hence, it is crucial for the industry to rely on a free real-time, low-latency, and decentralized settlement system [7].

IOTA combines both Fog and Mist into a new distributed computing solution. This can be seen as a combination of smart sensors with built-in computational capabilities (mist computing) with nearby processing stations (fog computing). IOTA micro-transactions enable party A's sensor data to be processed by Party B's processors in real time. In return, Party B can use IOTA to use resources from Party A or any other technological resource from other parties [8].

In this new autonomous Machine Economy, IOTA can be viewed as its backbone. The Tangle[3] ledger is able to settle transactions with zero fees so devices can trade exact amounts of resources on-demand, as well as store data from sensors and data loggers securely and verified on the ledger [9].

IOTA is different from the approaches taken by Bitcoin and Ethereum.[4] One main differentiating factor is that IOTA does not use blockchain, but instead it uses "Tangle," a Directed Acyclic Graph shaping up a Tangle. Secondly, blockchain is not suited to support micro-payments. Conversely, Tangle supports micro-payments by enabling IOTA to be efficient, scalable, and lightweight. But rather than being isolated paradigms, both approaches can work together. IOTA can communicate to

[2]Mist computing decreases latency and increases subsystems' autonomy. This takes fog computing concepts further by pushing some of the computation to the very edge of the network, to the sensor and actuator devices that make up the network [6].

[3]The main innovation behind IOTA is the Tangle, a novel new blockless distributed ledger which is scalable, lightweight and for the first time ever makes it possible to transfer value without any fees. Contrary to today's blockchains, consensus is no-longer decoupled but instead an intrinsic part of the system, leading to decentralized and self-regulating peer-to-peer network [9].

[4]Ethereum is a decentralized platform that runs smart contracts: applications that run exactly as programmed without any possibility of downtime, censorship, fraud or third party interference [10].

blockchain, that allows a future collaboration between IoT and established digital currencies. In fact, the IOTA project could even be used as an oracle to complete smart contracts [11].

4.2.4 Autonomous Decentralized Peer-to-Peer Telemetry

The exponential growth that IoT is experiencing is making increasingly important to have decentralized networks as a means to eliminate single point-of-failures that are associated with traditional centralized networks, as a way to increase its robustness and reduce the infrastructure and maintenance costs to manufacturers and vendors. By using the devices themselves as computational, storage, and communication nodes, we can build "hybrid" IoT systems where the "edge" complements centralized systems. We argue that edge computing will become a frontier of new economic value, creating an Economy of Things. As a way to spark adoption and change the current *status quo*, IBM and Samsung have developed the Autonomous Decentralized Peer-to-Peer Telemetry (ADEPT) proof-of-concept [12].

With it, they were able to demonstrate a distributed systems capable of sustaining a fully decentralized framework for IoT. As its backbone, ADEPT uses the blockchain to build a decentralized and distributed network of things [13, 14], using a combination of proof-of-work [15] and proof-of-stake [16] to secure transactions. This work was supported by using three distinct protocols:

- **BitTorrent**—BitTorrent is used to the file sharing.
- **Ethereum**—Ethereum is necessary to understand smart contracts and capabilities. At this point, the blockchain comes into the process.
- **TeleHash**—TeleHash is used to make the peer-to-peer messaging, since it is designed to be decentralized and secure, it fits in this system [17].

As a proof-of-concept for ADEPT, researches have deployed these three protocols that into a commercial washing machine (Samsung W9000) that was programmed to work with the ADEPT system, making an "Autonomous Washing Machine Orders Detergent" [18]. The goal is to automate the process of the ordering supplies. This process makes use of smart contracts to define the commands to receive a new batch of supplies. This way, the device can order and pay by itself when the capacity of the detergent is low. This payment is made using the blockchain. Later, the retailer receives the notice that the detergent has been paid for and ships it. Moreover, the owner of the washer can also be notified of the purchase details in its smartphone via its home network.

Another use case consists in a decentralized advertising marketplace using Large Format Displays (LFD)s to share and publish content without a centralized controller. The concept consists in an LFD, or more commonly a conventional display, where we can share the screen with anybody.

We have to choose the LFDs where the advertising will be published and also choose the advertisements (video files served by BitTorrent) to be published. Then, the advertiser receives the request through peer-to-peer messaging by TeleHash. After this, the content is shared and published. Finally, the advertiser receives the analytics, confirms the approval, and finalizes the payment.

4.3 Multi-Party Computation

After the era of connecting places and connecting people, the Internet of the future will also connect things. These "things" have sensitive information and data that can be shared but requires privacy. Secure multi-party computing is a technique that can be used here, since its purpose is to have multiple parties exchanging secret information privately without the need of a trusted third party. More formally, MPC consists of two or more parties, where each party has their own secret input. MPC computes some joint function f, that receives as input the secret information of each party.

It can be better explained with one of its well-known use cases, commonly referred as the millionaire's problem. Assuming that we have three parties: Alice, Bob, and Charlie. Each party uses respective inputs x, y, and z denoting their salaries. The goal is to find the highest salary of the three, without revealing their respective salaries. Mathematically, this can be achieved by computing:

$$f(x, y, z) = max(x, y, z)$$

Each party will share his secret input without reveal. At the end of the protocol, each participant will get only the result of the function f, without getting anything else about the other parties input, i.e., the secret inputs will not be revealed. The security of such protocols is defined with respect to the ideal model where f is computed by a trusted party T_f. During the execution of a protocol, the parties cannot get information about the inputs of the other parties. A third party *Tf* computes a function f receiving the inputs from the parties and after this, computes f, and finally sends the output back to the parties.

MPC based on secret sharing refers to methods for distributing a secret for a group of participants. Each participant has a share of the secret. The secret can be reconstructed only when a sufficient number of shares are combined together, where a threshold cryptosystem $(t + 1, n)$ where n is the number of parties and $t+1$ is the minimal number of parties to decrypt a secret encrypted with threshold encryption. We can see an image that illustrates MPC in Fig. 4.1.

A real-world example for MPC's applicability can be one where a patient wants to access his clinical records. He can make use of his private DNA code to make a query to a medical database of DNA related diseases. However, the patient does not want the hospital, and potential others, to know his DNA and health status. At the same time, the hospital does not want to disclose its entire DNA database to the patient. This is a problem, where the privacy must be preserved and can be solved while using MPC.

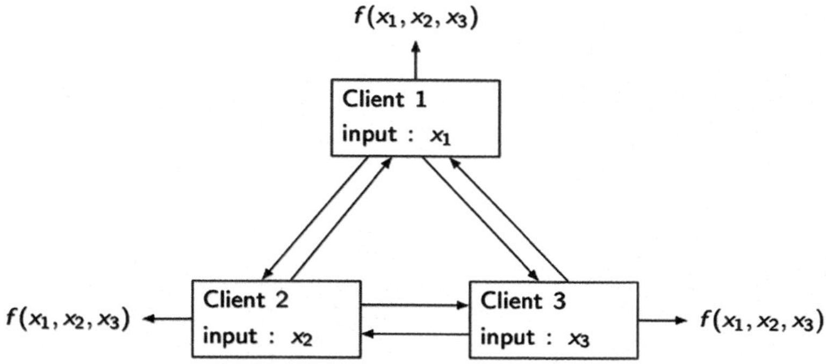

Fig. 4.1 Secure MPC without a trusted third party

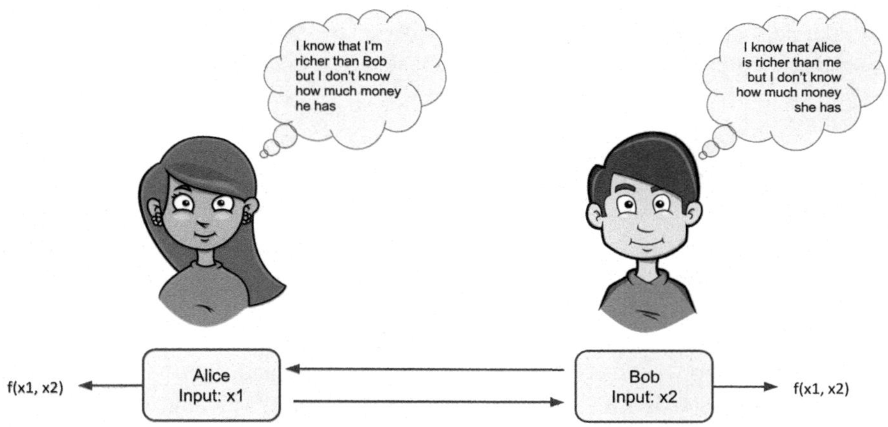

Fig. 4.2 Millionaire's problem

Extending our previous discussion, the millionaire's problem was first introduced in 1982 by Yao [19]. Suppose that we have Alice and Bob, and they want to know who is richer without reveal to each other or to a trusted third party the amount of money that they have or any type of additional information. The function $f(x1, x2)$: if $x1 > x2 =$ Alice else $x1 < x2 =$ Bob computes the inputs and returns the name of the richest (we can see the example in Fig. 4.2). Alice knows that is richer than Bob, but does not know how much money he has, and Bob also knows that Alice is richer, but does not know how much money she has. Therefore, in this protocol the privacy of each data is preserved, since they never reveal their salary.

In [19], it also describes other potential applications of the multi-party computation such as secret voting. It consists in a number of m members having globally to decide on a yes–no action. Each member has to choose an option xi and the result is computed by the function $f(x1, x2, x3, \ldots, xm)$. In turn, this function gives the final result without disclose the opinion of any other members and thus preserving privacy.

Another possible application is related with oblivious negotiation. In this case, we have Alice trying to sell Bob a house, each one having a strategy of negotiation in mind. Alice has possible strategies numbered as $A1, A2, \ldots, At$ and the same for Bob as $B1, B2, \ldots, Bu$. The result (no deal or sell at x dollars,...) will be decided once the actual strategies Ai, Bj used have been determined. The result is wrote as $f(i,j)$. This way, it is possible to carry out the negotiation obliviously, since Alice will not gain any information on Bob's negotiation tactics, expecting that it is consistent with the outcome, and vice-versa.

The last problem presented in [19] focuses on privately querying a database. Suppose that Alice wants to compute a function $f(i,j)$ and Bob $g(i,j) =$ constant. Bob does not know anything about i in the end. If we assume Bob as a database query system, with j being the state associated with the database, Alice can perform a query with the number i, and then, she can get an answer without getting any other information besides the data strictly required by her query. Conversely, the database system does not know which element was queried by Alice. This allows for users to preserve their privacy while avoiding data leakage from the database system.

4.3.1 Framework Analysis

Frameworks provide a set of solutions to commonly known problems for specific application domains, and are normally supported by a set of libraries. Developers can reuse code provided in these libraries and avoid handling with domain specific problems or low-level coding techniques. In most cases, frameworks result from a collaborative effort, as such, the burden of maintaining and improving it relies on a community instead of a single individual. A community offers a crowd-sourcing mechanism that enables users and developers to obtain information and resources to overcome issues found.

Despite their intrinsic advantages, there also some disadvantages associated with them. Namely, creating a framework is difficult and time consuming, i.e., expensive, and the learning curve can be steep. Moreover, they often add up to the size of programs, a phenomenon termed "code bloat" [20]. Over time and depending on the number of features introduced, a framework can become increasingly complex.

This added complexity can surpass the gains obtained from using a framework and the intended reduction in overall development time may not be achieved [20]. However, if the know-how can be further re-used in future projects, then this learning curve can be considered as an investment as it can be amortized across multiple projects [21].

There are some frameworks designed and implemented that enable secure multi-party computation (SMPC) providing basic MPC functionality that allow algorithm designers to build complex applications. Different flavors can be found for security level, accessibility, software composition, usability, scalability, and performance. MPC frameworks allow users to specify a SMPC where a number of parties execute a cryptographic protocol to do some joint computation with an agreed function without leaking any information of their inputs. An example could be an election where the correct tally is computed without revealing any information on the individual votes. Using a framework, the protocol is run without the players revealing anything about their inputs. Follows the set of MPC frameworks that we have tested.

4.3.2 *Virtual Ideal Functionality Framework*

Virtual Ideal Functionality Framework (VIFF) is a framework that allows users to specify SMPC, and is implemented in *Python* using *Twisted* [22] Framework to manage communication and General Multiprecision PYthon (GMPY) project [23] more specifically, the GNU Multiple Precision Arithmetic Library for the precision arithmetic. This framework is able to run on any platform where *Python* runs such as, *Linux*, *Windows*, and *Mac OS X* [24]. Protocols implemented in VIFF can be compositions of basic primitives like addition and multiplication of secret-shared values, or one can implement new primitives. In short, the goal of VIFF is to provide a solid basis where practical applications using MPC can be built [25]. VIFF's features include:

- Arithmetic with shares from Zp or $GF(2^8)$.
- Secret sharing based on Shamir and pseudo-random secret sharing (PRSS).
- Secure addition, multiplication, and exclusive-or of shares.
- Comparison of secret-shared Zp inputs, with secret Zp or $GF(2^8)$ output.
- Automatic parallel (asynchronous) execution.
- Secure communication using SSL [26].

4.3.3 *Sharemind*

Sharemind is a framework for privacy-preserving computations. It consists in a computation runtime and associated programming library for creating private data processing applications. This enables users to develop and test their custom privacy-preserving algorithms. As a result, one can develop secure multi-party protocols without the explicit knowledge of all implementation details. This also allows developers to test and add their own protocols to the library, as Sharemind is an open-source project [27]. The experimental Sharemind SDK contains the *SecreC 2* programming language that separates public data and secrets on a type system level and an emulator that developers can use to estimate the running time of their applications in a fully secured environment. SecreC programs are fully compatible with the Sharemind Application Server, that provides full cryptographic protection and supports enterprise applications [28].

4.3.4 SPDZ

SPDZ implements a general multi-party computation protocol secure against an active adversary corrupting up to n-1 of n players [29].

The processing model implemented by SPDZ is as follows: (a) an offline phase, where some shared randomness is generated, but neither the function to be computed nor the inputs need be known, and; (b) an online phase, where the actual secure computation is performed.

In the latter, we have active security [30, 31] with the following associated feature set:

- It uses BDOZ/SPDZ style MACs.
- It uses the n-party variant of the Tiny OT protocol to perform the pre-processing (outlined in some of the papers below).
- Works over any finite field GF(p) for p bigger than 40 bits; which is needed for statistical security. In practice to support floating and fixed-point operations p may be 128 bits in size.
- Provides actively secure offline and online phases.
- It provides a python based front end to produce byte-code for execution by the system [32].

4.3.5 FairplayMP

Fairplay [33] is a full-fledged system that implements generic Secure Function Evaluation (SFE). SFE allows two parties to implement a joint computation, that in real-world applications may be implemented using a trusted party, but does digitally without any trusted party. However, the Fairplay system uses Yao's Garbled Circuits (GC) and only supports secure communication between two parties. FairplayMP was created as an extension that appears to counter this limitation and introduce multi-parties. The extension to the multi-party case is needed since cryptographic protocols for the multi-party scenario are completely different than protocols for the two-party case [34]. This version implements secure computation using Yao circuits and secret sharing techniques.

4.3.6 Secure Computation API

SCAPI is an open-source general library tailored for Secure Computation implementations. This framework provides a flexible and efficient infrastructure for the implementation of secure computation protocols. Moreover, it also provides uniformity by offering a modular code-base to be used as the standard library for Secure Computation. SCAPI is also efficient because it is built upon native C/C++

libraries using JNI. SCAPI tries to improve adoption by developers by proving a clean design, streamline source code, and detailed documentation [35].

4.3.7 TASTY

Tool for Automating efficient Secure Two-Party Computations (TASTY) is a tool that uses Homomorphic Encryption (HE) or Garbled Circuits (GC) or the combination of both for describing and generating efficient protocols for several privacy-preserving applications [36]. TASTYL, that is the programming language adopted and created by TASTY, is an intuitive high-level language for describing the SFE protocols as sequence of operations on encrypted data (based on GC and HE). Also, TASTY allows to automatically analyze, run, test, and benchmark the two-party SFE protocol.

4.3.8 SEPIA

Security through Private Information Aggregation is a Java library for generic SMPC [37, 38]. It is tailored for network security and monitoring applications where the basic operations are optimized for large numbers of parallel invocations. SEPIA's basic primitives are optimized for processing high-volume input data. It uses Shamir's secret sharing scheme and is secured in the honest-but-curious adversary model [39].

4.3.9 Comparison Between MPC Frameworks

In practical terms, a number of frameworks and specialized programming languages have been created to implement and run SMPC protocols (Table 4.1). Fairplay, FairplayMP, and TASTY were built on top of the idea of "Garbled Circuits (GC)."[5] Sharemind and SPDZ use additive secret sharing[6] over a ring. In the case of VIFF,

[5] Yao's GC, utilized for SMPC, allows multiple parties to compute an arbitrary Boolean function on their individual inputs without revealing information about those inputs to any trusted third party, as long as they are semi-honest [41, 42].

[6] "Additive sharing supports efficient addition and multiplication due to the algebraic properties of the scheme. However, floating-point arithmetic is much more sophisticated and contains a composition of different operations, both integer arithmetic and bitwise operations." [43].

Table 4.1 Comparison between MPC frameworks

Framework	Programming language	Techniques	Number of participants	Year of creation
VIFF [25, 26]	Python	Secret sharing	$\geqslant 3$	2007
Sharemind [27, 28]	SecreC (C++)	Secret sharing	3	2006
SPDZ [30–32]	Java/C++/Python	Secret sharing	$\geqslant 2$	2016
SCAPI [35, 40]	Java	GC	$\geqslant 2$	2013
FairPlay [33]	SFDL (Java)	GC	2	2003
FairPlayMP [34]	SFDL (Java)	GC and secret sharing	$\geqslant 3$	2006
TASTY	TASTYL (based on Python)	GC and HE	2	2009
SEPIA	Java	Secret sharing	$\geqslant 3$	2008

FairplayMP, and SEPIA, they were built on top of Shamir's secret sharing[7] [44, 45]. Lastly, TASTY uses combinations of GC and HE techniques [36].

The main application for GC is to secure two-party computation. For more than two parties, secret sharing schemes [46] are normally used. All these frameworks support a similar set of primitives including addition, multiplication, comparisons, and equality testing. Programming on these platforms either uses a specialized language, or a standard programming language and library calls, depending on the platform [45].

Some of these frameworks are more accessible for non-experience developers, more specifically, VIFF, Sharemind, SPDZ, FairplayMP, TASTY, and SEPIA. SPDZ has the particularity of having the online and offline phases built-in into the framework.

Adding examples to the source code in the frameworks is something that helps the development stage. For instance, if we start with a new API, sometimes we would not be capable of implement new examples because we don't know the structure of the entire framework. Sometimes, it would be more easier to study an implementation of a known standard protocol to understand the structure. In this case, an example could be the millionaire's problem.

In terms of known programming languages, VIFF, SPDZ, and SEPIA can be more easier to adapt. These frameworks use standard programming languages such as Python, Java, and C++, making them more accessible. Alternatively, TASTY and FairplayMP have the TASTYL and SFDL specification which may require more adaptation time.

[7]Shamir's Secret Sharing is a form of secret sharing, where a secret is divided into parts, giving each participant a random part of the secret, where some of the parts or all of them are needed in order to reconstruct the secret. Sometimes, it is used a threshold scheme to define k parts that are sufficient to reconstruct the original secret, since can be impractical to have all participants to combine the secret [44].

SCAPI is the preferable framework for advanced users, as SCAPI is an open-source general library tailored for Secure Computation implementations. It is best suited for users that already are knowledgeable on how the protocol works and therefore only need a library to implement secure protocols.

4.3.10 Future Research Directions

In this section, we discuss the main research questions and summarize the identified challenges in the MPC area. For this study, we analyze some papers that present future challenges.

Havron et al. [47] describe a problem that can be solved with MPC in the future. Social scientists and researchers are always the need to make data analysis. However, they have to reveal some of the input data to another party to perform the analysis. Often, they cannot make the data analysis due to legal restrictions and privacy issues. This can be solved by MPC for scientific analysis of large data. A new research pathway lays on the improvement of MPC implementations to enable novel scientific data methods through the creation of new tools that will make these techniques accessible to social scientists. This points to the need for a closer examination of automatic data-matching between separate datasets with private set intersection, improving fixed-point integer conversion for decimal data values used in computation, and other privacy-preserving applications. To summarize, the ultimate goal is to achieve this without disclosing private information, i.e., the inputs of each party.

Since the number of devices in IoT is growing, the data that is being exchanged is also increasing. For this, it would be important to have a filter in order to identify non-sensitive data, making tools that help us detect sensitive versus non-sensitive data.

The authors of the paper [48] propose an interesting research direction for "MPCs on Bitcoin" where Alice and Bob can determine who is the wealthiest one based on who has more coins. However, this is only possible if each party is interested in proving that it is the wealthiest one, because every participant can easily pretend to be poorer than it really is and "hide" its true wealth by transferring it to some other address under its control.

The authors of the paper suggest that analyzing what functionality can be computed this way, i.e., taking into account the problem of the participants may pretend to be poorer than they are, may be an interesting research direction. In our opinion, this can be a possible research direction not only in the "millionaire's problem" but also in other problems that are isomorphic in nature, just with varying underlying contexts. There are still open problems in the MPC such as constructing protocols that are secure against "malleability" and "eavesdropping" attacks [48].

Another issue is related with the information leakage from memory access patterns, that is solved by cryptographic techniques like oblivious RAM and Private Information Retrieval (PIR). Both techniques can be used in a black-box manner to resolve the aforementioned issues but the actual selection of a scheme and associated security and complexity analysis is still subject for future research. Arguably a more difficult challenge is the protection against misuse of branching instructions where proper code obfuscation seems a possible solution, in particular by hiding control flow operators such as "then" and "else." In this case, the code obscurity is a necessity to thwart chosen instruction attacks by stripping the semantics from the instructions (much like encryption removes the meaning from a plain-text by casting it into a cipher-text) [49].

4.4 Multi-Party Computation and Blockchain

With the ever-increasing pervasiveness of IoT, there is a necessity to create platforms that are both decentralized and private. The combination of SMPC with blockchain technology can be an important advance in this area, as it may be possible to create platforms that enable privacy-preserving (data remains encrypted even in-use) and are resilient. Although some issues remain unanswered, namely is it possible to design a decentralized platform without completely relying on a trusted third party, or can one construct a fully decentralized protocol to the sell secret information without allowing sellers and buyer to cheat?

4.4.1 Applications

Enigma [4] combines the use of SMPC and blockchain technologies. In the next sections, we describe Enigma and associated use cases with real-world applications.

4.4.2 Enigma

Enigma is a peer-to-peer network that enables different parties to jointly store and run computations on data while keeping the data completely private. This model works in parallel with an external blockchain technology. Similar to Bitcoin, Enigma removes the need for a trusted third party.

The main motivation for this work lays in the avoidance of centralized architectures that might lead to catastrophic leakage of data that would result in the loss of privacy. Their approach is designed to connect to an existing blockchain and off-load private and intensive computations to an off-chain network. Code is executed both on the blockchain (public tasks) and on Enigma (private and computationally intensive tasks). Opposing blockchain, which only ensures correctness in execution, Enigma is able to simultaneously provide privacy and correctness. One of its main features is its privacy-enforcing computation, as Enigma can execute code without data leakage while still ensuring correctness. Since heavy duty computations are a known issue for blockchains, Enigma avoids it by only allowing running computations to be broadcasted throughout the blockchain. While blockchains are not mean to be general-purpose databases they can be used to strategically store information. Enigma has a decentralized off-chain distributed hash-table that makes use of blockchains to store data references (not to the actual data). Nevertheless, private data must be encrypted on the client-side before storage and access-control protocols should be programmed into the blockchain, that will act as a public proof for the authorization schema.

4.4.3 Enigma Application Cases

SMPC can be applied in some fields where privacy is a concern. In this section we describe some of the most relevant domains where we envision it can be applied.

Applicability in IoT seems rather straightforward, since we can store, manage, and use highly sensitive data collected by IoT devices in a decentralized, trust-less cloud. The crypto bank is also a field where the internal details have to be anonymous, so we can run a full-service crypto bank without exposing information about its internal design and implementation. The autonomous control of the blockchain allows users to take loans, deposit cryptocurrencies, or buy investment products without publicly revealing their financial situation.

In line with the millionaire's problem, where n-parties want to know if they are more wealthier than the others, without exposing their financial status to each one, there is blind e-voting. In the latter case, not only the privacy of each voter is maintained but also the actual vote count can potentially remain private.

Another application of Enigma is on the N-factor Authentication, where voice, face, and fingerprint recognition are all stored and computed on Enigma. As the access-control is supported by private contracts, only the user is able to access its own data.

Furthermore, private contracts are used when we want to securely share some data with a third party. We can define some policies in the contracts which restrict the access to data, maintaining and enforcing control and ownership. The shared data on MPC is always reversible, since third parties do not have access to actual raw data, and are restricted to only running secure computation over it. The identity management is also supported by private contracts, since when an user wants to log in, an authenticating private contract is executed in order to validate the user and to link to his real identity with a public pseudo identity making this process completely trust-less and privacy-preserving. This way, the authentication and store identities are fully anonymous, and the user on Enigma only has to secret-share her personal information required for authentication.

For data protection, privacy-preserving approaches should be paramount to companies, since they hold large volumes of potential sensitive user data that is a potential target for criminals. With Enigma, companies can use data to provide personalized services and match individual preferences without storing or processing the data on their servers, and thus removing the security and privacy risks. By doing so, companies are also protected against corporate espionage and rogue employees. It should be noted that employees can still use and analyze data for the benefit of the user while enforcing agreed consents. With these solutions in place, companies can potentially provide access to the data while preserving security and privacy.

A potential interesting case can be found in the data marketplace, for example, a pharmaceutical company looking for patients for clinical trials can scan genomic databases for candidates. In this process, consumers can sell access to their data with guaranteed privacy, autonomous control, and increased security. The marketplace would eliminate tremendous amounts of friction between companies and individuals, lower costs for customer acquisition, and offer a new income stream for consumers.

4.4.4 *Future Research Directions*

Enigma has yet to release any source code, as so, its off chain-network performance is still unknown. However, historically, computation on encrypted data has been slow in practice, and it remains an active research path [50].

Additionally, Enigma entails techniques for processing transactions without knowing their contents that might provide an alternative way to achieve similar accountability benefits while supporting transactions. However, this approach does not preclude the possibility of a validator favoring transactions based on a bias, because it can identify the transactions with help of colluding peers, even if the transactions are encrypted [51].

Enigma's novel combination of three paradigms, secret-sharing, MPC, and P2P, opens new possibilities to address current open issues on data privacy and the growing liabilities faced by organizations that store or work on large amounts of personal data. However, until the official launch of Enigma's source code it is not possible to guess what problems will be solved by it [52].

4.5 Summary

This chapter has provided a summary of the multiple concepts of the secure computation in different approaches. The first Sects. 4.2 and 4.3 presented some concepts of secure computation such as secure multi-party computation that consists in exchange data anonymously without a trusted third party, or blockchain that is a secured way of online transaction. We described the concepts and its applications, as well as the combination of both (if any) and/or with Internet of Things. In the secure multi-party computation section, we describe the frameworks that we tested, as a way of analyzing the functionality of each one.

With this analysis, we discovered some interesting applications which show that there are some attempts developments based on the combination of some of these concepts. The main finding applications were:

- Blockchain with IoT: IOTA and ADEPT (Sect. 4.2.2);
- Blockchain with secure multi-party computation: Enigma MITs (Sect. 4.4.1).

As we describe in Sect. 4.2.1, there are several limitations in order to integrate the blockchain technology with IoT. Section 4.2.2 describes a solution that consists of using fog computing as a way of decreasing latency, having "local awareness" and mobility (including vehicular mobility). As the limitations of cloud computing are undesirable for IoT, the integration between IoT and blockchain should be made using the fog computing with IoT.

However, there are unsolved problems and open issues yet, that we described as future research directions in Sects. 4.3.10 and 4.4.4.

References

1. D. Ron, A. Shamir, Quantitative analysis of the full bitcoin transaction graph, in *International Conference on Financial Cryptography and Data Security* (Springer, Berlin, 2013)
2. M. Swan, *Blockchain: Blueprint for a New Economy* (O'Reilly Media, Sebastopol, 2015)
3. A. Dorri, S.S. Kanhere, R. Jurdak, Blockchain in internet of things: challenges and solutions (2016). arXiv preprint arXiv:1608.05187
4. G. Zyskind, O. Nathan, A. Pentland, Enigma: decentralized computation platform with guaranteed privacy (2015). arXiv preprint arXiv:1506.03471
5. What is IOTA? https://iota.readme.io/v1.1.0/docs. Cited 23 January 2017
6. J.S. Preden et al., The benefits of self-awareness and attention in fog and mist computing. IEEE Comput. Mag. **48**, 37–45 (2015)
7. M. Atzori, Blockchain-based architectures for the internet of things: a survey. Browser Download This Paper (2016)
8. IOTA: Economy of Internet-of-Things (2016). https://medium.com/\spacefactor\@mDavidSonstebo/iota-97592581f985#.rhosuii7l. Cited 10 November 2016
9. IOTA (2016). http://www.iotatoken.com/. Cited 10 November 2016
10. Ethereum - Homestead Release Blockchain App Platform (2013). https://www.ethereum.org/. Cited 11 November 2016
11. IOTA: Internet of Things Without the Blockchain? (2016) http://bitcoinist.net/iota-internet-things-without-blockchain/. Cited 10 November 2016
12. P. Veena et al., Empowering the edge-practical insights on a decentralized internet of things. IBM Institute for Business Value **17** (2015)
13. Autonomous Decentralized Peer-to-Peer Telemetry (2015). http://wiki.p2pfoundation.net/Autonomous_Decentralized_Peer-to-Peer_Telemetry. Cited 11 November 2016
14. M. Signorini, Towards an internet of trust: issues and solutions for identification and authentication in the internet of things. Ph.D Thesis, Universitat Pompeu Fabra (2015)
15. S. Nakamoto, Bitcoin: a peer-to-peer electronic cash system (2008). https://bitcoin.org/en/. Cited 18 April 2017
16. S. King, S. Nadal, Ppcoin: peer-to-peer crypto-currency with proof-of-stake. Self-published paper (2012)
17. TeleHash - Encrypted Mesh Protocol (2014). http://telehash.org/. Cited 11 November 2016

18. IBM & Samsung live demo of ADEPT — TheProtocol.TV (2015). https://www.youtube.com/watch?v=U1XOPIqyP7A. Cited 11 November 2016
19. A.C. Yao, Protocols for secure computations, in *23rd Annual Symposium on Foundations of Computer Science, 1982. SFCS'08* (IEEE, New York, 1982)
20. N.M. Edwin, Software frameworks, architectural and design patterns. J. Softw. Eng. Appl. **7**(8), 670 (2014)
21. I.P. Vuksanovic, B. Sudarevic, Use of web application frameworks in the development of small applications, in *MIPRO, 2011 Proceedings of the 34th International Convention* (IEEE, New York, 2011)
22. Twisted Matrix Labs: Building the engine of your Internet (2016). http://twistedmatrix.com/trac/. Cited 25 October 2016
23. The General Multiprecision PYthon project (GMPY) (2008). https://wiki.python.org/moin/GmPy. Cited 25 October 2016
24. A. Aly, Network flow problems with secure multiparty computation. Diss. Ph.D Thesis, Universté catholique de Louvain, IMMAQ (2015)
25. I. Damgård et al., Asynchronous multiparty computation: theory and implementation, in *International Workshop on Public Key Cryptography* (Springer, Berlin, 2009)
26. VIFF, the Virtual Ideal Functionality Framework (2007). http://viff.dk/. Cited 25 October 2016
27. D. Bogdanov, S. Laur, J. Willemson, Sharemind: a framework for fast privacy-preserving computations. in *European Symposium on Research in Computer Security* (Springer, Berlin, 2008)
28. Sharemind SDK Beta (2015). https://sharemind-sdk.github.io/.Cited25October2016
29. I. Damgård et al., Multiparty computation from somewhat homomorphic encryption, in *Advances in Cryptology—CRYPTO 2012* (Springer, Berlin, 2012), pp. 643–662
30. Y. Lindell et al., Efficient constant round multi-party computation combining BMR and SPDZ. in *Annual Cryptology Conference* (Springer, Berlin, 2015)
31. I. Damgård et al., Practical covertly secure mpc for dishonest majority—or: Breaking the spdz limits, in *European Symposium on Research in Computer Security* (Springer, Berlin, 2013)
32. SPDZ Software (2016). https://www.cs.bris.ac.uk/Research/CryptographySecurity/SPDZ/. Cited 25 October 2016
33. D. Malkhi et al., Fairplay-secure two-party computation system, in *USENIX Security Symposium*, vol. 4 (2004)
34. A. Ben-David, N. Nisan, B. Pinkas, FairplayMP: a system for secure multi-party computation, in *Proceedings of the 15th ACM Conference on Computer and Communications Security* (ACM, New York, 2008)
35. SCAPI Documentation (2014). https://scapi.readthedocs.io/en/latest/intro.html. Cited 30 October 2016
36. W. Henecka et al., TASTY: tool for automating secure two-party computations, in *Proceedings of the 17th ACM Conference on Computer and Communications Security* (ACM, New York, 2010)
37. M. Burkhart et al., Sepia: security through private information aggregation (2009). arXiv preprint arXiv:0903.4258
38. M. Burkhart, M. Strasser, D. Many, X.A. Dimitropoulos, SEPIA: privacy-preserving aggregation of multi-domain network events and statistics, in *USENIX Security Symposium, USENIX Association (2010)*, pp. 223–240
39. SEPIA - Security through Private Information Aggregation (2011). http://sepia.ee.ethz.ch/. Cited 10 November 2016
40. Y. Ejgenberg et al., SCAPI: the secure computation application programming interface. IACR Cryptol. **2012**, 629 (2012). ePrint Archive
41. P. Chen, S. Narayanan, J. Shen, *Using Secure MPC to Play Games*. (Massachusetts Institute of Technology, 2015)
42. A.C.-C. Yao, How to generate and exchange secrets, in *27th Annual Symposium on Foundations of Computer Science, 1986* (IEEE, New York, 1986)

43. P. Pullonen, S. Siim, Combining secret sharing and garbled circuits for efficient private IEEE 754 floating-point computations, in *International Conference on Financial Cryptography and Data Security* (Springer, Berlin, 2015)
44. A. Shamir, How to share a secret. Commun. ACM **22**(11), 612–613 (1979)
45. K.V. Jónsson, G. Kreitz, M. Uddin, Secure multi-party sorting and applications. IACR Cryptol. **2011**, 122 (2011). ePrint Archive
46. Y. Huang et al., Faster secure two-party computation using garbled circuits, in *USENIX Security Symposium*, vol. 201(1) (2011)
47. S. Havron, Poster: secure multi-party computation as a tool for privacy-preserving data analysis (University of Virginia, 2016)
48. M. Andrychowicz et al., Secure multiparty computations on bitcoin, in *2014 IEEE Symposium on Security and Privacy* (IEEE, New York, 2014)
49. S. Rass, P. Schartner, M. Brodbeck, Private function evaluation by local two-party computation. EURASIP J. Inform. Secur. **2015**(1), 1–11 (2015)
50. B. Yuan, W. Lin, C. McDonnell, Blockchains and electronic health records(2015). http://mcdonnell.mit.edu/blockchain_ehr.pdf. Cited 18 April 2017
51. M. Herlihy, M. Moir, Enhancing accountability and trust in distributed ledgers (2016). arXiv preprint arXiv:1606.07490
52. Blockchain and Health IT: Algorithms, Privacy, and Data (2016). White Paper

Chapter 5
Self-Aware Fog Computing in Private and Secure Spheres

Kalle Tammemäe, Axel Jantsch, Alar Kuusik, Jürgo-Sören Preden, and Enn Õunapuu

5.1 Introduction

As our capacity for monitoring and measuring objects, activities, and processes is growing exponentially, we also find many new applications in our immediate environments, our bodies, and our homes. Wearable sensors for measuring our leisure and sports activities as well as our health conditions have proliferated and gained acceptance. Already today many of us continuously carry around several sensing, computing, and communicating devices wherever we go. Also, our homes are becoming increasingly smart as they are equipped with cameras, motion detectors, and environmental sensors that provide data for air conditioning controllers, surveillance, and medical monitoring. The advances of sensing, computation, and communication technology are also being utilized in military applications. While the military solutions used to represent the cutting edge of technology, with the rapid advancement of computing and sensing technology civilian applications are showing the way for future solutions.

The focal topic of the book—where and how the data processing should be performed in future systems—is a very relevant one for the private spaces of our bodies and our homes due to three main considerations: efficiency, privacy, and dependability. The alternatives to be considered are between the two extremes of

K. Tammemäe (✉) • A. Kuusik • E. Õunapuu
Tallinn University of Technology, Tallinn, Estonia
e-mail: kalle.tammemae@ttu.ee; alar.kuusik@ttu.ee; enn.ounapuu@ttu.ee

A. Jantsch
Institute of Computer Technology, TU Wien, Vienna, Austria
e-mail: axel.jantsch@tuwien.ac.at

J.-S. Preden
Thinnect, Inc., Sunnyvale, CA, USA
e-mail: jurgo@thinnect.com

A.M. Rahmani et al. (eds.), *Fog Computing in the Internet of Things*,
DOI 10.1007/978-3-319-57639-8_5

fully centralized and fully distributed processing. All the sensed data can be sent to a cloud server, recorded, archived, processed, and used for making decisions that are then either returned to an actuator close to the sensor or sent to another appropriate agent like a hospital. At the other end of the scale, the sensor nodes themselves could do almost all the data analysis and decision-making and only selected, abstracted data is sent to outside agents such as the hospital which is necessary to realize a required function.

Transferring data to the cloud for processing takes time and resources, and the delays are not always acceptable in latency sensitive applications like health-monitoring, emergency-response, etc. [1]. The solution is to introduce a hierarchy where time-sensitive processing is done at the edge of the network. This is where Fog computing emerged along with IoT, with the role to mediate resource-abundant and slow cloud computing and agile but only partially informed edge computing. Edge computing can offer especially low latency response by providing limited computing resources at the very edge of the network. A more capable variant of Edge Computing, "Mist computing," offers the same benefits with more flexibility and manageability—with Mist computing nodes forming dynamic partnerships for data exchange and execute complex tasks requested by other Mist, Fog, or Cloud nodes. Due to added resilience against communication instabilities, the Mist computing is often discussed in connection of outdoor applications, e.g., connected vehicles, intelligence surveillance, etc.

Cloud computing can be also a source of data privacy concerns. The privacy of individuals is easily protected if the sensitive data does not leave the sensor, the body, or the house where it originates. Once the data has left the private sphere, we need sophisticated, complex, and expensive technical, institutional, and legal solutions to ensure a robust protection of privacy. However, if locally processed data does not leave the private sphere, it is relatively easy to install mechanisms to guarantee that data is transmitted to outside agents only with explicit permission of the owner. For example, we have used wearables with accelerometer and pulse rate measurement to differentiate between different indoor activities/behaviors with a good success rate for decision-making about the physical health of the individual [2]. Still, it is an information which is not expected to leave the private sphere.

The growing concern in the IoT era is the energy consumption. Even mostly in low power mode, the 14 billion network enabled devices, which are currently in use worldwide, waste 400 TWh (terawatt-hour) of electricity per year. With 50 billion devices in 2020, the consumption is expected to increase by a factor 3.5 to over 1400 TWh [3]. Since communication over the network requires energy, the overhead of sending data to a cloud server for processing is proportional to the amount of data and the distance. On the other hand, processing data locally requires more complex and costly nodes. This implies a complex trade-off, in particular if local nodes are powered by battery or harvested energy. While the specific shape of the trade-off curves depends on the application details, we can identify the main factors. First, the energy for local processing has to be compared with the energy for communication plus the energy for cloud processing. Second, as will be explored below under the term "attention," the possibility to tune the amount of sensed, communicated, and processed data to the actual needs in a particular situation of the system can save significant energy by avoiding unnecessary activities. Local processing has the advantage to be able to react faster to changing needs due to the omission of communication delays.

In fact, it can be argued that sophisticated local processing not only can lead to reduced energy consumption but also to improved adaptation and more robust behavior. Self-awareness holds the promise that the local node or subsystem has a comprehensive understanding of its situation and its environment which leads to better decisions on what data to collect, how to process it, what data to communicate, and what decisions to take. Self-awareness implies [4, 5]:

- that a device can assess the quality of the sensed data, i.e., it keeps track of precision, accuracy, and completeness of the collected data;
- that the system understands its own performance, i.e., if it has performed well or badly in recent past and over a longer time period; and
- that the system understands if the environment is meeting its expectations, e.g., if it is in fact in an environment in which its operation is meaningful.

An active and quickly growing area of research [6–10] explores the possibilities, costs, and implications of self-awareness in various application domains. In this chapter, we will focus on the base of case studies how self-awareness can be realized in these application domains and its potential benefits.

The paper is organized as follows: The cloud, fog, and mist computing in various application areas is examined in Sect. 5.2. In Sect. 5.3, a self-aware data processing at fog and edge nodes is discussed. There are four related case studies—health monitoring in Sect. 5.4, home patient safety monitoring and training support described in Sect. 5.5, household self- and remote control in Sect. 5.6, and intelligence surveillance in Sect. 5.7. The analysis of fog-level smart gateway implementation feasibility and options is discussed in Sect. 5.8. The final Sect. 5.9 offers concluding remarks.

5.2 Cloud, Fog, and Mist Computing Networks

Taking a step back and analyzing why we deploy sensing systems, we realize that the reason for this is to collect data for decision-making. The decisions could be made by a machine or a human, based on the type of the decision that must be made, the types and amount of data being collected, and the data processing methods that must be selected. This selection process is part of the data to decision paradigm, introduced by Broome [11].

With the data to decision paradigm, the data collection is driven by the decision-making processes; from the data and information perspective, it does not make any difference whether the data is being processed in the device that collects it or whether it is sent to the cloud for processing. However, many of the

Table 5.1 Selected attributes of mist, fog, and cloud computing

	Abstraction	Distance (m)	Delay (processing + communication) (s)	Bandwidth (b/s)	Power (W)
Mist	Operational	10^{-2}	10^{-6}	10^{3}	10^{-2}
Fog	Functional	10^{2}	10^{-3}	10^{6}	10^{2}
Cloud	Conceptual	10^{6}	10^{0}	10^{8}	10^{5}

While Cloud computing can provide the highest (conceptual) level decisions, the limited resources of Mist computing are sufficient only for simple operational control. The Fog capabilities fall in between allowing adaptive (functional level) decision-making. The figures give only an approximate order of magnitude

functional and nonfunctional system parameters are affected by the selection of the computing architecture when choosing between Cloud, Fog, and Mist computing architectures as seen in Table 5.1. Some of these parameters include latency of control loops, bandwidth usage, storage requirements, security and privacy aspects, system robustness, and reliability. In applications where low latency from the sensor to the data consumer (which can be an actuator) is critical, Mist Computing architecture is beneficial, which is also the case when potentially large amounts of data are collected, which could be processed locally. Fog Computing architecture provides value when data from multiple sensors need to be fused and the resulting information be provided to a consumer locally, which may be the case, for example, when the heating or cooling requirement for a building needs to be determined based on the occupancy level of all the rooms in the building.

Cloud computing is applicable when we are either processing big amounts of data, for which processing methods may be complex (e.g., data collected over long period of time, when the amounts of data are too big to be processed by edge devices or when the processing methods are too complex to be executed on edge devices or when the data sources are so distant that collecting data to a central location is feasible, for example, when behavioral data is collected from a city population).

Fog computing brings computation closer to the edge of the network. In Fog computing, a more capable device (e.g., the gateway) bears the responsibility for data processing or IoT application execution, regardless whether the application is just simple data collection or building automation with many actuation tasks. Placing the application logic in a gateway has many advantages, such as simplicity of coordination (the centralized control paradigm used with Fog computing is very similar to conventional programming paradigms), simple management of application logic (the application logic is all concentrated in a single device), and having access to macro-level information (e.g., house or city block) in the application from all sensors. However, this approach also has drawbacks, such as increased delays in applications involving control and unnecessary high bandwidth requirements as all data must move through the gateway. The gateway is a single point of failure for applications that must be executed on the network and the operation of the entire network is dependent on the gateway.

Mist computing takes Fog computing concepts even further by pushing appropriate computation to the very edge of the network, to the sensor and actuator devices that make up the network. With Mist computing, the computation is performed in the microcontrollers of the sensor or actuator nodes. The Mist computing paradigm decreases latency as devices are able to communicate directly with each other (making data directly available to the consumer) and further increases the autonomy of a solution.

When designing a solution using Mist Computing principles, a monolithic architecture with dedicated device roles and interaction patterns is not feasible as it severely limits the applicability of the solution. To create a solution, which can be deployed in variable configurations and that adapts to the changes of configurations, a service-based architecture is optimal. By applying the principles of a service-based architecture, the application can be described as a combination of services, which are dependent on each other. Any device that has access to the network can subscribe to a service that is offered by any of the devices on the network. Hence, in a temperature control application a heating unit can subscribe to temperature information from all the temperature sensors that are located in the room, which the heating unit is heating. The sensors can start providing their temperature data directly to the heating unit using the interval specified by the heating unit (the interval being dependent on the properties of the room and the power of the heating unit, and the relation of the parameters can be determined automatically at run time by the heater). No human involvement is therefore needed for setting up the application, simplifying network configuration.

Unlike a system with a fixed structure, where the functionality of the components and their interaction patterns are well controlled and predictable, in a dynamic Mist Computing scenario the interactions are not fixed as the system configuration itself is not fixed during design time. In the context of nondeterministic interactions between systems, the temporal and spatial validity of the data that is exchanged is crucial for ensuring correctness of the outputs of algorithms using the data as an input.

Therefore, each device in the network must be aware of its location as most applications tend to be location dependent. The necessary "location awareness" can be created at installation time (by "telling" the device its location), or the devices can determine their location autonomously by determining their location relative to some existing beacons, with known locations (for example, a light fixture in a room may be aware of the room where it is located and all other devices in the room can determine their location based on proximity to the light fixture). Also the devices must share a common clock or there must be a way of temporal alignment of data to ensure temporal validity of data used in computations.

The services provided by end devices may also be requested by mobile devices or servers, in which case the service request reaching a specific network is routed to the device, which is able to provide the specific service. This means that in one network end devices and a gateway may be both providing services to the same server. As an example in a building automation scenario we may be interested in room occupancy information for every single room, so all the occupancy sensors in the individual rooms must report information directly to the server, while the operation times of standalone AC units may be aggregated (in the gateway) to estimate the total power usage of AC units in the building.

5.3 Self-Aware Data Processing

The umbrella term *self-awareness* encloses a number of concepts such as self-adaptation, self-organization, self-healing, self-expression, and other self-* properties. Different authors endow these terms with different, only partially overlapping meanings, but probably most will agree that self-awareness in computing devices holds the promise that those devices exhibit more sensible behavior under novel conditions and adapt more gracefully to faults, failures, and changing environments. Ultimately, a self-aware system should fully understand its own situation and detect its own misbehavior or underperformance due to:

- faults, that may be caused by aging, accidents, or a physical attack,
- a malicious attack on its functions, or
- functional design errors in its hardware or software.

After deviations are detected by the self-awareness monitor, appropriate actions can be taken by the parts of the system that are typically considered outside the self-awareness monitor proper. Such actions may range from raising an alarm to an abrupt stop of all operations, or be a less extreme adaptation of behavior.

Recently, many projects have been initiated to fulfill parts of this promise. Due to the breadth and versatility of the term, its application in a wide range of different domains with various objectives and assumptions has been reported under diverse labels such as autonomic [12], nature-inspired [13], organic [14], self-organizing [15], self-adapting [16], cognitive [17], or self-aware [18] computing.

Since there is no widespread consensus on the meaning of these terms, we review briefly the properties that we consider to be part of self-awareness [4, 19, 20].

Semantic Interpretation includes an appropriate abstraction of the primary input data and a disambiguation of possible interpretations and an assessment of the reliability of the data [5].

Desirability Scale provides a uniform goodness-scale for the assessment of all observed properties.

Semantic Attribution maps properties into the desirability scale suggesting how good or bad an observation is for the system.

Attention determines which data should be collected and analyzed, given limited resources [2].

History of a Property Awareness of a property implies awareness of its change over time.

Goals provide the context in which interpretation and semantic attribution is meaningful.

The Purpose of a smart embedded systems is to achieve all its goals.

Expectation on Environment The system expects a specific environment and detects if the environment deviates significantly from expectations.

Expectation on Subject Similarly, the system's own state and condition are continuously assessed to detect deviations, degradation, performance, and malfunctions.

The Inspection Engine continuously monitors and assesses the situation and integrates all observations into a single, consistent worldview.

Recalling the objectives of fog computing, namely efficiency of, privacy, and reliability, it turns out that self-awareness is apt to play a useful role. Sophisticated assessment of the system's objectives, resources, and needs will facilitate all three goals. First, self-awareness and overall efficiency follow similar trajectories in the design space. The more complete and correct the self-assessment is, the more effective will be the usage of resources with respect to given goals. A full understanding can lead to minimize the amount of computation and communication necessary close to the theoretically possible. Second, if privacy is acknowledged as an important objective by the system, the self-awareness component can track it and prevent unwarranted information leakage. And finally, maximizing local intelligence in the form of self-awareness makes the local system more independent and resilient against disturbances in the wider system.

Hence, it can well be argued that self-awareness is a potentially significant asset in a fog computing setting, but the specific trade-offs are sensitive to the constraints and requirements of the application.

5.4 Case Study I: Health Monitoring

The combination of progress in sensor technology and data analytics facilitates ever more sophisticated monitoring of vital signals for medical, professional sport, or leisure purposes. Inexpensive sensors for heart rate, blood pressure, respiratory rate, temperature, blood oxygen saturation, and many other parameters can be attached to the body for inferring activities, health conditions, fitness levels, and the efficiency of workouts and training sessions. As a multibillion Euro market for vital sign monitoring with double-digit growth rates emerges [21], plenty of investment money is poured into the sector with the consequence that more versatile and less expensive sensor devices and monitoring equipment will become available in the forthcoming years. Thus, data from vital sign sensors are plentiful but must be processed quickly and efficiently.

Table 5.2 Early warning scores derived from [25]

Score	3	2	1	0	1	2	3
Heart rate (bpm)	<40	40–51	51–60	60–100	100–110	110–129	>129
Systolic BP (mmHg)	<70	70–81	81–101	101–149	149–169	169–179	>179
Breath rate (per min)		<9		9–14	14–20	20–29	>29
SPO_2 (%)	<85	85–90	90–95	>95			
Body temp. (°C)	<28	28–32	32–35	35–38		38–39.5	>39.5

5.4.1 Early Warning Score

Consider the Early Warning Score (EWS) [22, 23] system, which is a manual tool widely used in hospitals to track the condition of patients. It allows for the evaluation of risks early in order to take preemptive actions and can be defined as "a specific procedure for the early detection of any departure from normal frequencies of clinical cases or serological reactors of specific diseases by monitoring a sample of the population at risk" [24]. Based on five physiological parameters, as listed in Table 5.2, it assigns a score between 0 and 3 to each of them, with a lower score meaning better condition. Adding up individual scores gives the EWS score between 0 and 15, which has been demonstrated to be a reasonable predictor for subsequent health deterioration and even mortality [24].

In current hospital practice, EWS is a manual procedure, but recently attempts have been made to mechanize the measurement and the EWS calculation based on wearable sensors [26]. This would have the significant advantage that the procedure is not bound to the hospital any more since trained medical personnel does not have to be present for performing the measurements. Continuous monitoring of patients at home or at work becomes feasible. Anzanpour et al. have demonstrated an automated EWS system with wearable sensors, a gateway node that relays the sensor data to a server, which in turn computes the score and makes the assessment. The server can be located in the hospital and medical professionals can further analyze the data and take actions as required. The advantages of this system are decreased costs, increased comfort for the patient, and increased monitoring coverage outside hospitals. The main disadvantages are the inconvenience of carrying sensors and being wired up, and a potential loss of quality in the measurements, due to incorrect attachment, loose contacts, and faulty sensors and equipment. To address these concerns, Götzinger et al. [27] have added the capability to analyze the data reliability and consistency thus making automated EWS more robust and reliable. Indeed, a general issue with the proliferation sensors connected to the IoT in a growing number of applications and domains is the unknown quality of the collected data. Deployed sensors exhibit a wide range of accuracy and precision, hardware faults and finite battery capacity limit their lifetime, and the hardware and software of the processing and communication equipment may have their own flaws and limitations. For these reasons, it is mandatory that a system like the automated EWS analyzes the quality of the collected data and keeps track of meta-information.

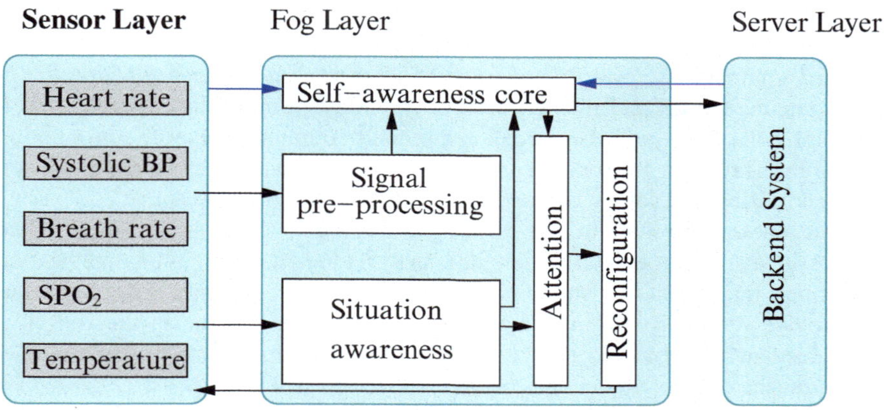

Fig. 5.1 Architecture of the automated early warning score (EWS) system [25]

To this end, Götzinger et al. describe an agent based processing system that assesses the consistency and the plausibility of the sensor samples by a set of interacting agents that specialize on several tasks: abstraction, history tracking, confidence derivation, and binding [27]. They demonstrate that several typical failure conditions in the data collection and processing chain can be correctly identified to improve the overall robustness of the EWS system.

Anzanpour et al. take this approach a step further and propose an architecture that combines self-awareness, situation awareness, an attention mechanism, and an adaptive resource allocation scheme, illustrated in Fig. 5.1. Based on the sensory input and the system's expectation on itself and its environment, a model of its own performance and relevant aspects of the environment are built up. For a correct computation of the EWS, the sensory data is preprocessed with traditional signal processing algorithms to suppress noise and extract relevant features. The preprocessing stage offers abstractions relevant to the medical application of the raw data to the upper layers in the processing chain. The self-awareness and situation awareness blocks in Fig. 5.1 adjust the primary input data in two ways. First, an ambiguity and consistency analysis identifies potential errors in the input data. Second, the activity of the patient is estimated, since the data interpretation is dependent on the activity pattern. The system distinguished five different activity modes: sleeping, resting, walking, jogging, and running. Hence, taking both measurement incorrectness and patient activity modes into account, the most like interpretation of the data is derived and the corresponding EWS value is computed.

Furthermore, the built-in attention mechanism allows for efficient resource usage without compromising the core objective which is a correct assessment of the patient's health. The system considers four levels of severity of the patient's condition (normal, low, medium, and high), five states for the activity (sleeping, resting, walking, jogging, and running), and four different situations for the environment (indoor night, indoor day, outdoor night, and outdoor day). Depending on the position in this three-dimensional space, priorities are assigned to the activities of the system and resources are allocated accordingly. The attention block in Fig. 5.1 computes the priority vector which is then used by the reconfiguration to configure the sensors and allocate resources properly. Its main objectives are, first, to maximize the quality of assessment, and second, to minimize power consumption. Possible parameters for sensor configuration are sampling rate, sample precision, bias and calibration values, and activity modes like sleeping and active.

The backend system in Fig. 5.1 might include a cloud server and medical professionals. It can do further detailed analysis, initiate or fine-tune medication, or request the patient to visit the hospital. For us most interesting is the possibility to provide feedback information to the self-awareness module in the fog layer. This feedback information can be the basis for learning, adaption, and improved performance of the system.

5.4.2 Discussion

The automated EWS system as described in [25–27] illustrates the benefits and trade-offs of a sensitive and informed decision-making in the fog layer. A self-awareness module as described above can increase the robustness of the data analysis and the quality of the assessment of the system's own situation, the monitored subjects state, and the environmental conditions. This in turn is a solid basis for correct decisions regarding the collection of further information, triggering emergency actions and a prudent usage of resources. Obviously, all the described functions can be realized in the server layer on a main server or a cloud computing infrastructure. However, several aspects have to be considered in this trade-off.

Energy efficiency By no means is it obvious which alternative is more energy efficient. Gathering a lot of processing and intelligence locally in the fog layer can reduce communication costs between sensors and the cloud server by orders of magnitude. However, computation may be more efficient in a high performance server infrastructure because of an optimized processor architecture, large caches, and deployment of the latest processor family. Also, if the fog layer runs on battery, a fraction of the energy is lost in nonideal batteries. On the other hand, local processing allows for customization by adjusting to the actual precision needed, by avoiding unnecessary generality in the architecture, and by eliminating unneeded computation altogether. Hence, we face a complex trade-off and the most energy efficient solution may be somewhere between the extremes of all-local and all-central computation and depend heavily on the specifics of the application.

Latency Likewise, the latency and response times depend on a variety of factors. To begin with, not every application is particularly latency sensitive. If it is, it makes a difference if average or worst-case latency is more important. On one hand, fog-level computation avoids the delays of communication between the local node and the server, but on the other hand, the server may be significantly faster in doing the required computation. Hence, again, the optimal solution with respect to latency is application dependent and may be located between the extreme points of the design space.

Customization of the hardware architecture and software can lead to significant gains in terms of optimality and efficiency. The downside is that customization requires effort in design, validation, and maintenance. Hence, the optimal point cannot be located in general terms, depends on the application, and will in most cases end up somewhere between the extremes.

Location of control If most of the processing is performed in the fog layer, it has two implications which concern efficiency, reliability, and privacy: First, the locally generated data never leaves the local environment, such as the person's private home. Usually, only abstracted data is sent to the main server and only in specific cases a full record is requested, for instance, when the case is unclear and a physician wishes to examine the details of the situation. Second, local configuration options, such as which data to sample and to store, are decided locally and not by the remote server. Locality of data and decision is an efficiency issue for the reasons discussed above under the terms energy efficiency and latency. It is a reliability issue, because even mundane and simple operations like counting the heart beat are dependent on a flawless internet connection and server infrastructure, which makes the system vulnerable to many kinds of disturbances. Finally, it is a privacy issue because only the fog-layer alternative provides the option that the user can control what data and decision leaves his/her private sphere. When data and configuration access rights are sent and stored at a server, it requires more elaborate and stringent policies to protect privacy concerns.

5.5 Case Study II: Patient Safety Monitoring and Training Support

The cost of hospitalization and elderly social care is increasing worldwide. Telecare is believed to compensate the reduced traditional clinical interactions and home nursing. IoT technologies certainly have a role in traditional telemedicine of chronic disease management as well as guaranteeing patients' safety at home, enabling new services like technology assisted rehabilitation. The market research company Gartner expects that healthcare related technologies and services will have 16 percent of IoT market business value in 2017.

Distributed and mobile sensing devices are well suitable for improving safety of elderly and special needs patients' at home. According to Centers for Disease Control and Prevention, about 800000 patients are hospitalized every year in the USA due to a fall down [28]. Ibid., 25% of older adults will fall every year in the USA. In addition, it has been shown that daily activities and cognitive and physical health might be in good correlation [29, 30].

Certain modern home telecare systems already provide fall down detection functionality based on wearable motion sensors, camera systems, and floor sensors [31]. Technologically it is also possible today to predict the frailty of elderly people based on activity of daily living (ADL) pattern analysis [32]. However, compared to conventional telecare solutions that just periodically transmit vital signs measurement data, the real-time and dependability requirements for such safety critical telecare solutions are certainly higher. By conventional home telecare systems, the measurements are usually conducted twice a day and even by ECG signal measurements the recording package does not exceed 50 kB. Therefore traditional central server or cloud based data store for personal health records (PHRs) has been sufficient so far. The technological and also privacy requirements for novel home telecare solutions employing certain motion capture or continuous ADL monitoring capabilities are significantly stronger which leads to use of alternative—distributed data processing architectures. It can be estimated that the raw data stream of an inertial measurement unit (IMU) capable for sufficiently accurate human activity or free fall tracking is at least 1 kbps. Similar average data stream can be expected from low frame rate safety observation cameras. It is simple to understand benefits of distributed fog-like data processing for such real-time patient safety monitoring solutions over the centralized ones. The remote server load and communication channel throughput can be significantly reduced through the local data aggregation and decision-making. Also, from the clinical point of view, only the aggregated meta-information, i.e., the number of active hours, mean activity level, sleep quality, and presence of special events like fall downs have significant long-term value worth for preservation in PHR. The rest of raw data has a low clinical value and raises the privacy concerns if the data is transmitted outside of the private areas.

Nowadays the smartphones are frequently used as telemedicine gateways and wireless communication is widely used. For the majority of sensor signals, i.e., temperature, conductance, movement and position, and illumination, the wireless transmission consumes 100–1000 times more energy than processing that is also motivating the local data aggregation. Distributed fog computing also increases the dependability of wirelessly networked systems. Today Bluetooth Low Energy, ANT+, and different IEEE802.15.4 standard compliant radios are mainly used for the personal area networking. Due to the throughput limits, the real-time data streaming may seriously affect the reliability of the communication. Fog computing reduces such risks as well because of relaxed requirements for the communication channel throughput. Even more, if the real-time critical data processing is done locally, redundant communication channels, i.e., over wireless MESH network can be effectively used. A typical distributed ADL safety monitoring system is shown in Fig. 5.2.

Theoretically it would be possible to acquire required user activity, location, and falling information from one wearable IMU device and process the data within the same sensor device. Modern IMUs usually have built-in free fall event detection. Combined use of linear accelerometer, gyroscope, and magnetometer data should be sufficient for dead reckoning based movement monitoring. In practice, sensor fusion with alternative smart home sensors still has to be used because of internal

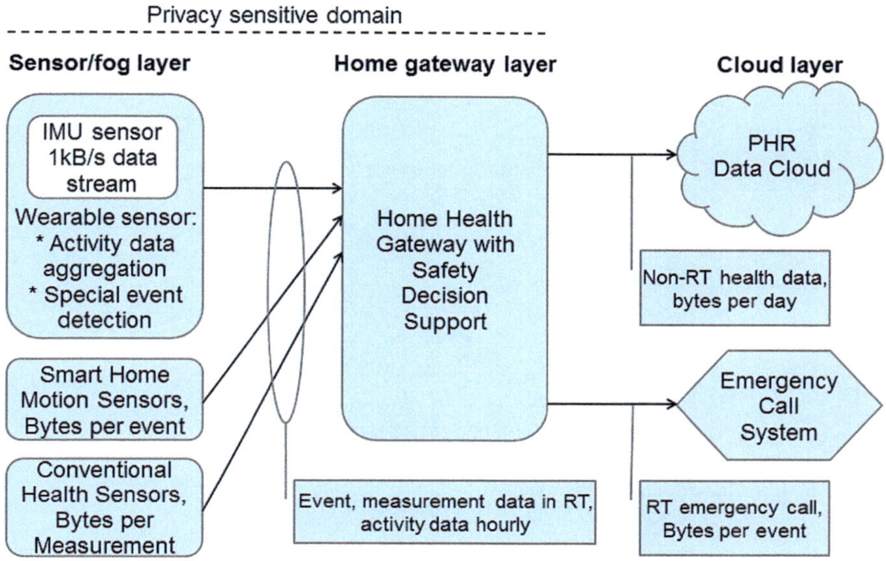

Fig. 5.2 Activity of daily living (ADL) safety monitoring and estimated data rates of various I/O channels

errors of inertial motion sensors. In real life, human fall down cannot be sufficiently reliably detected with inertial sensors only and due to nonlinearities of IMUs the dead reckoning movement tracking is reliable just for some meters. Due to the unpredictable network delays, the described sensor data fusion cannot be made at a remote location. Therefore, the sensor data fusion, fog-based aggregation, and possible reasoning at the gateway device is the most appropriate solution for intelligent home telecare systems supporting ADL analysis and hazard detection.

Reablement through the support to physical activities extends the independent living time and therefore reduces the needs for expensive traditional social care. For example, it has been reported cumulative cost savings of 30% in 2–5 years through the trainings [33]. However, reablement process itself that includes human assisted trainings and validation of new skills is costly and time consuming. It is expected that wearable sensors and other IoT devices will enable teletrainings and safety validation of home-based exercising.

Remote validation of physical training exercises through the home telecare systems may be possible in the near further. Such technological solutions would significantly reduce the needs for physiotherapists and clinical visits. For example, specific home exercising is required during the stroke recovery [34] and after joint replacement surgeries [35]. In both cases, rather simple exercises have to be performed to avoid irreversible processes of joint stiffening. For the training efficacy, certain exercising speed and amplitude have to be preserved which was not possible earlier. Today the training process may be well monitored with wearable

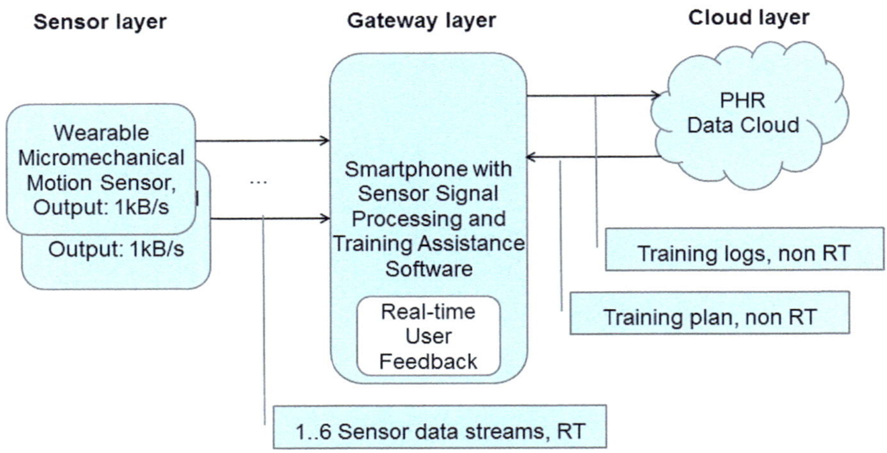

Fig. 5.3 Home-based training assistance

IoT devices giving real-time feedback to the user if the exercising is accurate or not. It would be logical to connect such training assistance unit with the above described home telecare system, which will effectively enable the targeted machine-assisted exercising. Essentially the system should transmit the quantity and quality of performed training exercises to the PHR cloud server where the clinicians and physiotherapists can access the data and make further treatment decisions as seen in Fig. 5.3. As in a previous ADL monitoring example, it is feasible to process the sensor data locally in the fog to minimize the load of communication channels and save resources of PHR repository server. In this particular case, the decision about exercising correctness has to be essentially made on-site. Local decision support is required to provide exercising feedback in real time, without noticeable latency, and to meet safety critical dependability needs of the full data path. The most practical is to implement training assessment processes in the gateway device which usually has sufficient amount of the computing power and can access to the context information regarding the environment and user. Fog-based local decision support also preserves the user privacy because it minimizes the personal information amount to be transmitted to the remote locations.

5.6 Case Study III: Smart House

In the chapter, we analyze the experience and lessons learned from the multiyear use of a smart house solution. The focus is on smart house energy management (HEM).

Main principles used in implementation of this smart house solution are following:

- Computations are made up close to the sensors and actuators, both for security as well as the latency reasons. Home gateway (a fog computing node) is the place where they are held. The Cloud computing is used for analyzing big amount of data and for decision models developing and testing.
- Self-Awareness is widely used. System adapts to the specific sensed situation. The system is very cautious with respect to non-authorized objects. Process of the authorization of object is two-step—at first, the system administrator has to introduce the device to the system, and only after that the device will be added to the system. The system continuously scans the devices accepted and in case of mismatch refuses to interact with the device.

Abowd and Day [36] introduced the primary context types:

- Location—location data from the GPS sensor,
- Identity—identify object based on the RFID tag,
- Time—read time from clock, also daytime, and
- Activity—what activities are in progress?

There is also a secondary context type, which is derivative information that we can use based on the primary context. For example, using Identity, we can obtain considerable information about a person in social networks and the internet.

In our case, the context information is the place where we can begin our analysis. Our system has ability to react to the new situations and learn from results. For our approach, the adaptive and self-organizing properties are exceptionally valuable. Instead of attempting to find the most optimal solution based on the available information, we attempt to use self-organization methods. For example, it is known that house energy planning is a highly complex and difficult problem. The situation can change notably quickly, and our perfect plan fails. Instead of careful planning, we attempt to use adaptive techniques. For example, we can obtain information about energy needs from the sensors and quickly use stored energy (Tesla Powerwall).

Before we go to real examples, let's look at the concept of the smart solution.

The smart-solution concept is notably important in the context of this example. Smart refers to quick intelligence. People are considered smart when they show up a rapid ability to respond intelligently to different situations. As we observe, smartness is strongly connected with the concept of intelligence. It is a long debate regarding whether we can exhibit the intelligence to computers or software. Today's computers can do many things that require intelligence, such as driving a car off-road or on city streets.

The term "intelligent systems" is used to describe the necessary level of capabilities to achieve the system goals. Intelligence has been scientifically categorized as a biological stimuli response mechanism. In our case, we obtain the stimulus from the environment using different sensors and make a response using the knowledge that we have and the actuators that are connected to the system. During its lifecycle, the system learns from experience. The learning ability is precisely what makes the system intelligent. Computer power and the amount of information and sensors make a system smart.

Smart solutions are composed of smart objects [37]: *"One definition of smart objects is a purely technical definition–a smart object is an item equipped with a form of sensor or actuator, a tiny microprocessor, a communication device, and a power source. The sensor or actuator gives the smart object the ability to interact with the physical world. The microprocessor enables the smart object to transform the data captured from the sensors, albeit at a limited speed and at limited complexity. The communication device enables the smart object to communicate its sensor readings to the outside world and receive input from other smart objects. The power source provides the electrical energy for the smart object to do its work."* These objects can learn and adapt to different real-world situations, and different machine learning algorithms are used [38].

5.6.1 Case Study Object Characterization

This is a six-member family living house, where electricity use is high as seen in Fig. 5.4. The system has two heat pumps, gasoline power generator, and three oil radiators. The largest consumers of electricity are water heater, washing machine, and stove. The house usage is irregular because the residents are working adults, and some days they do not use the house. All this makes it difficult to optimize energy use but gives a great economic effect.

Traditional approach to make house energy systems is to define system requirements, designing solution, and implement solution by professional's team. All this is costly and needs time to get results. Typically, the end user will not be able to actively influence the system behavior and functionality. In our approach, we choose another path.

1. We enable intelligent decisions by liberating device data across operations.
2. Decisions are defined by user using rules.
3. System collects decision data together with context data, makes analysis of collected information and makes recommendations.

One example that can be used to optimize the use of energy is the heating device duty time. The aim is that upon returning home, the family has a house comfortably warm, while the heaters are not turned on too early. The house settings user interface[1] is presented in Fig. 5.5. The interface has both informational role (room temperature, light level, and humidity) as well as the characteristics of actuator triggering role. For example, if the user clicks a camera icon, he/she gets the camera video stream or in the case of clicking a plug icon the light or heat pump will be toggled.

[1]Telia Eesti AS https://www.telia.eeSmartHomesolution. The service marketing has been discontinued since 2017.

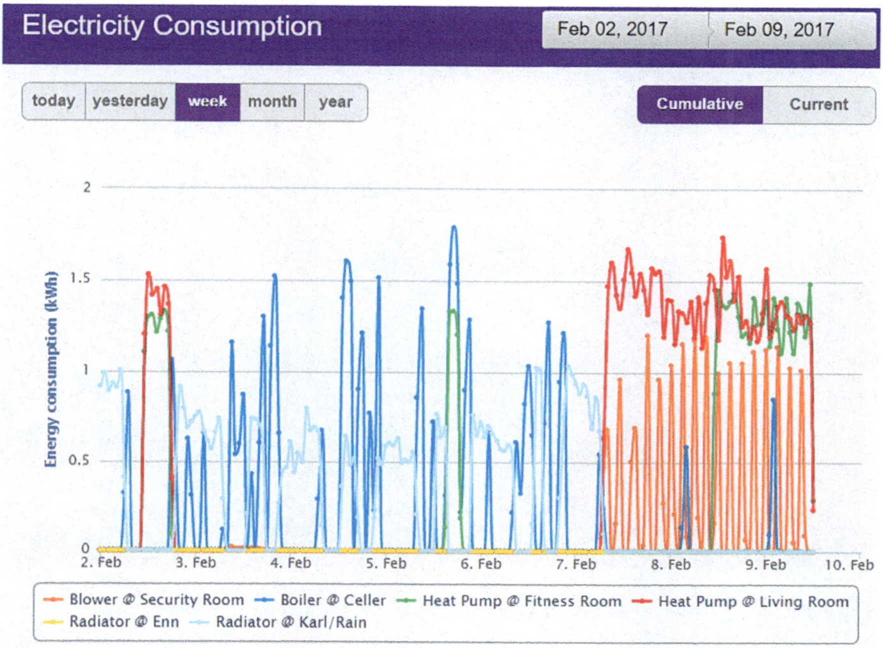

Fig. 5.4 Typical weekly electricity consumption in house

Users of the smart house are very satisfied with the solution due to a number of reasons:

- Security. The house has a security system to detect and alarm in case of smoke/fire. There are day/night surveillance cameras to monitor the household in habitats absence.
- Convenience. The house itself (heating and airing) is continuously tuned according to user needs.
- Money savings. Controlled heat pumps and optimized usage of the water heater reduce costs.

We plan to complement the solution to the solar panels, energy storage, and energy purchase from the market. It is also planned to set up a small-scale local energy grid.

In conclusion, it can be said that the principles set in this section turned out to be fruitful. Decisions should be taken as close as possible to the equipment following the fog computing paradigm and using as much as possible the context information. Only in case of large volumes of data, the advantages of cloud computing should be exploited.

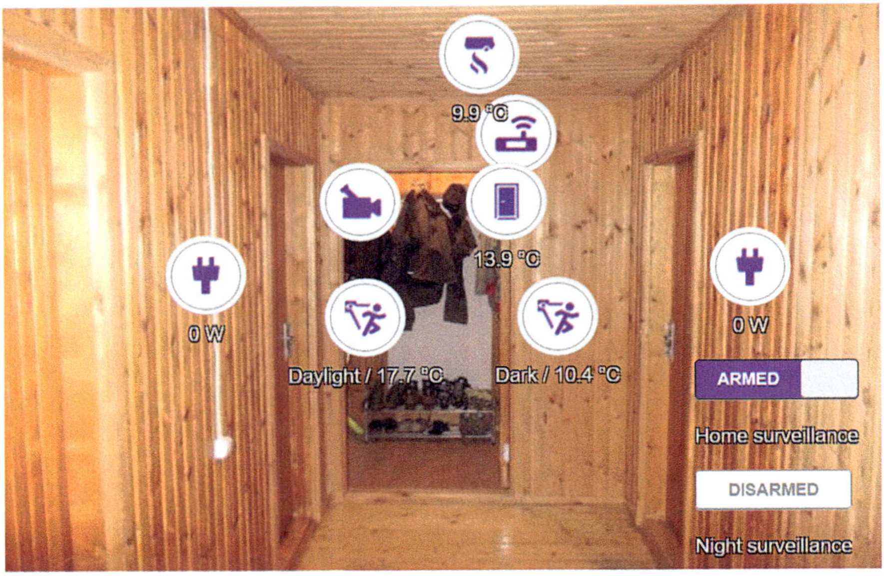

Fig. 5.5 Family living house settings user interface

5.7 Case Study IV: Intelligence Surveillance, Reconnaissance: Military Sensing Systems

As stated in the beginning of the chapter, military sensing systems are also undergoing an architectural change, driven by the enhanced capabilities of the system components. While Network Enabled Capabilities (NEC) was a compelling vision in the very beginning of the century, we can say in 2016 that the technology components and architectures are catching up with that vision. This section describes an Intelligence, Surveillance, and Reconnaissance (ISR) solution concept, which builds upon the concepts of Mist and Fog Computing and that is very close to the NEC vision introduced more than a decade ago.

In the European Defense Agency's IN4STARS project, the Research Laboratory for Proactive Technologies developed an ISR solution prototype, which relied on Mist and Fog computing methods. The task of the solution was to provide enhanced situation awareness for units in the field and to remote intelligence operatives. The sensor system deployed in the field processes collected data locally at the sensor level using Mist and Fog Computing principles and delivers only situational data and information that has been requested by the users to the users, following the Data to Decision [11] paradigm.

Unlike the classical system approaches that assume a central coordinating agent, the sensor system architecture applied in the project builds upon a mixed Fog and Mist computing approach, where the individual Mist Computing nodes are autonomous. When a request for information is made to a deployed ISR system, any node that is capable to provide the requested information with an acceptable cost will respond to it. The specific sensor modalities needed for providing the requested information (e.g., detection and identification of tracked vehicles) need not be colocated with the system providing the information, instead the information may be fused from several sources, including both ground based as well as airborne sources. To enable this kind of operation, the nodes must maintain a certain level of self-awareness as well as awareness of the system itself, in order to find the required sensor sources for generating the information requested by the information consumer. In order to achieve and maintain the required self- and group awareness, the individual systems must be able to communicate directly and to request services from other systems. The conceptual system configuration is depicted in Figs. 5.6, 5.7, and 5.8.

Applying the D2D approach in a Fog Computing paradigm means that the requests for situational information made by the information consumer can be directed to the sensor assets in the field, closest to the area of interest. The routing of

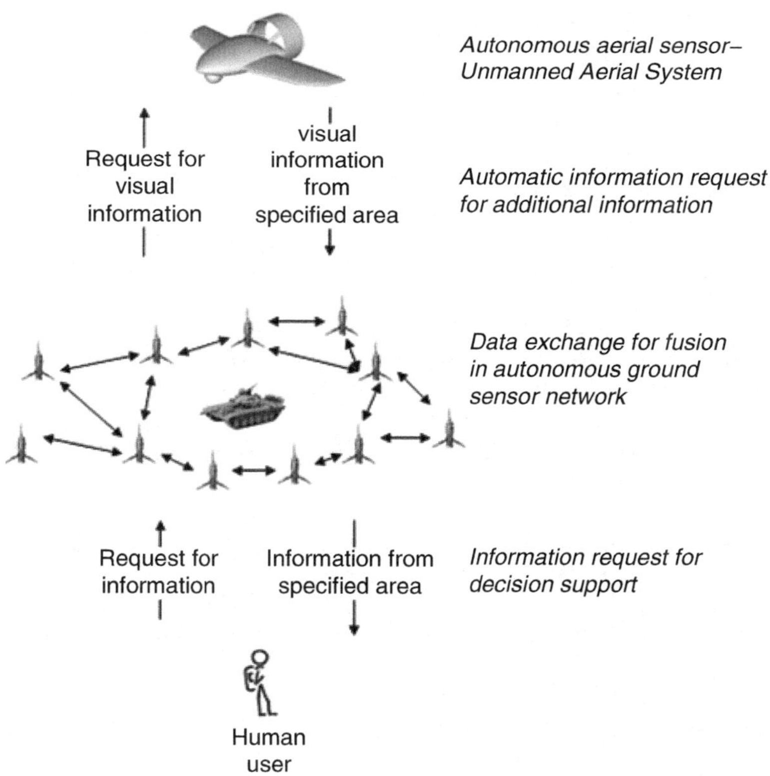

Fig. 5.6 Data flows in an ISR solution making use of Fog and Mist Computing principles

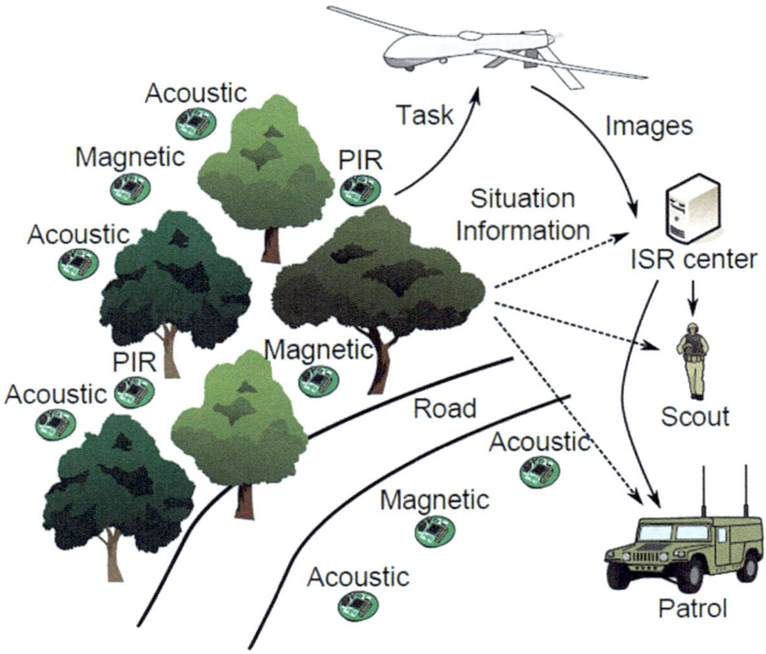

Fig. 5.7 Sensor layout of an ISR solution

information to the specific information provider may be done using many alternative methods, e.g., geo-routing, using a central service directory or some other service discovery mechanism. Based on the information requests, the algorithms are primed in the computing device providing the information service (e.g., sensor or fusion node). Service requests are made to the data sources (sensor nodes) from which data is needed for computing the requested situational information. Once the information has been computed, it is provided to the consumer.

The sensor system was built as a wireless sensor network with a dynamic network structure and functionality and ad hoc communication paths. Depending on the information request received by the system, the appropriate Mist and Fog Computing algorithms are triggered locally to process the collected sensor data and to deliver the requested information to the user. Every sensor that was part of the solution was equipped with a local computation unit and a wireless communication interface, making it an independent node in a Mist Computing solution. The multimodal sensors (seismic, infrared, acoustic, visual, magnetic, etc.) employed in the solution made use of in-sensor signal processing, including novel detection and classification algorithms based on in-depth analysis of sensor data. The solution comprised several sensors, which were assembled into a system in an ad hoc manner, enabling real-time configuration and behavior adaptation.

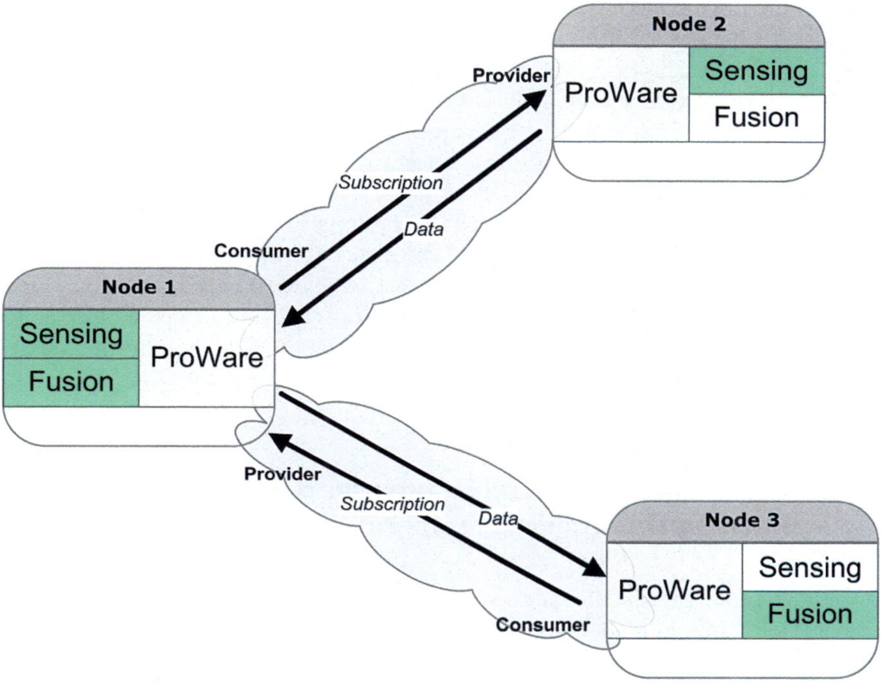

Fig. 5.8 Subscription (service) based data exchange between edge devices in Mist Computing

In order to assess the locations and types of the detected objects with a higher precision and to service the situational information needs of users, a Fog Computing layer was added, which performed in-network distributed data aggregation and fusion, which relied on a service-oriented middleware based on a subscription model. At the Fog layer, data fusion methods were applied, with on-demand data to sensors, collecting data needed for a given fusion operation based on the information requests that had been made by the user. This logical system structure enabled hierarchical buildup of situation information by distributed fusion and aggregation. In order to achieve correctness of data, time selective communication was used to provide temporally aligned data for fusion and aggregation algorithms at the Fog layer.

The distributed processing implemented using Fog and Mist Computing principles was enabled by the Proactive Middleware, which has been developed at the Research Laboratory for Proactive Technologies at Tallinn University of Technology. ProWare offers the services of data provider discovery, online data validation, and service contract agreements between data providers and consumers. Such an architecture facilitates predictable operation also in a changing system configuration. With these features, ProWare enables dynamic creation of data and information exchange relationships in a distributed network using a subscriber model. ProWare solves one of the major challenges in a distributed computing scenario, which is ensuring the temporal and spatial validity of data—making sure that the data used in computations originates from the right location (i.e., from the correct sensor node) and is temporally coherent with other data used in the computation. The latter is very critical in sensor data fusion applications, but difficult to achieve when the fusion is performed in a distributed manner using Mist and Fog principles.

The sensor systems that made up the ISR prototype can be categorized to ground sensors and airborne sensors, below both types are described and the operation of the systems discussed. The object localization solution based on acoustic arrays utilizes autonomous acoustic arrays working together for localizing detected objects. The same arrays can be also used for acoustic classification using any of the available classification methods as we have also presented in our previous work [39].

5.8 Requirements and Architecture for a Smart Gateway Based on Hierarchical Temporal Memory

Common in all case studies above is the need for an intelligent processing unit capable of learning, model building, behavior adaptation, and anomaly detection. The backpropagation learning algorithm in traditional NNs is slow [40] and requires intensive use of floating point arithmetic (e.g., to calculate sigmoid function), which makes it ill-suited for applications where "human-like" quick reaction and continuous learning ability is expected.

Among contemporary continuous learning algorithms, the Hierarchical Temporal Memory (HTM) of Numenta, inspired by the architecture of the neocortex, has proved to be successful in comparison with many other detector algorithms [41]. Moreover, the Numenta approach is attractive due to a smaller computational load where instead of complex floating point calculations simple fixed point and integer arithmetics is used. This property makes HTM attractive to be tested in application areas where memory, processing performance, and energy are limited.

In this section, we discuss an approach for the design of HTM based micro-controller and SoC computing platforms for fog computing gateways, to meet the required processing capabilities, memory, and energy consumption.

As illustrated in Fig. 5.9, we consider gateways responsible for the following functions and actions:

- Edge input devices control and power management;
- Input data fusion and time-stamping;
- Data abstraction and conditioning;
- Encrypted data exchange with cloud service;
- Data encoding to Sparse Data Representation format for HTM processing (spatial and temporal pooling of data);
- Continuous/dynamic model building, prediction, and anomaly detection;

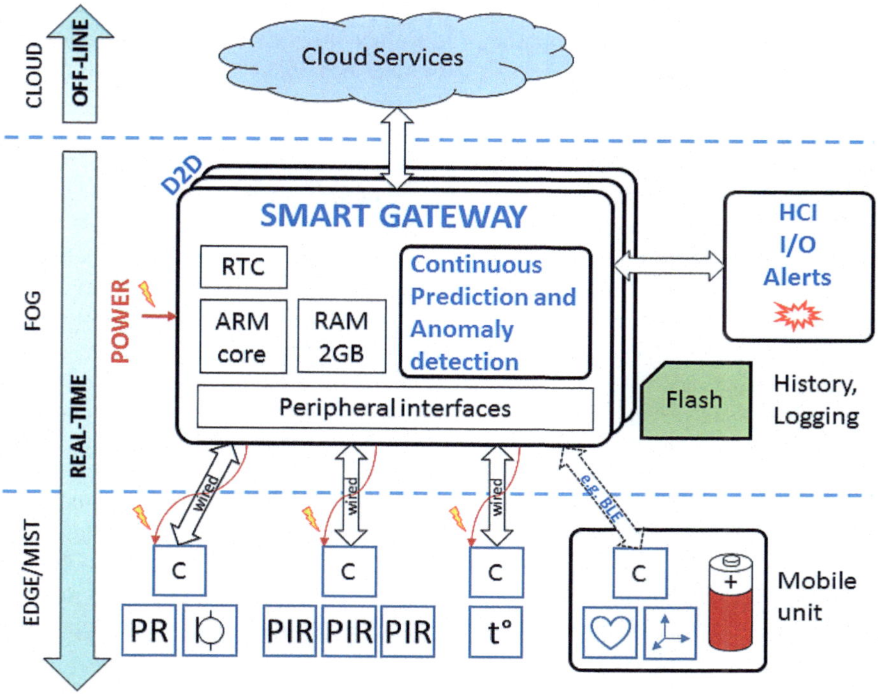

Fig. 5.9 Smart gateway according to the fog computing paradigm. *C* controller, *PIR* Passive Infrared Sensor, *PR* Photo-resistor, *D2D* Device-to-Device communication, *HCI* Human–Computer Interface

– The first level reasoning and decision-making about the situation;
– User feedback and interaction (Human–Computer Interface), and
– Introspection: The service quality over time and the history of actions.

Microcontroller/microcomputer platforms might be capable for aggregating sensory data and running the HTM (Hierarchical Temporal Memory) based data processing (prediction and anomaly detection). As an example of mobile platforms, even the ARM microprocessor based smartphones could be viable solutions due to great connectivity and abundance of memory. Still, the online massive computations required by HTM might be largely impeded due to multiple regular applications of the smartphones. Their energy reserve cannot guarantee 24/7 readiness and in addition, continuously online devices pose a risk to confidentiality and privacy [42]. Due to that stationary, single-board computers like Raspberry Pi (https://www.raspberrypi.org/), Pine 64 (https://www.pine64.org/), or Odroid C2 (http://www.hardkernel.com/) might be preferred.

Currently, the HTM tool NuPIC can be compiled to run on desktop or laptop platforms, and there are limited attempts to get it working on single-board computers like Raspberry Pi 3 or on Android platforms, which are actually the most attractive gateway modules in Fog computing due to their low price, high availability, and good community support. As a result, NuPIC is available for RasPi 3 platform only since June 2016 [43].

Despite mentioned above positive properties, the pure software implementation of HTM is resource demanding. Although the basic operations are simple, the need for memory and performance in real-time situations is high (GigaBytes of memory, multicore processors, etc.). Due to that nature, the HTM is not applicable at far edges of the network but it is rather a gateway level of functionality to manipulate higher levels of real-time data which is already preprocessed, filtered, averaged, and fused from raw data at edge sensor nodes. Still, there is no good comparative study of energetic aspects of different NN and HTM algorithms.

HTM functionality equipped gateway might be used to learn and assess the patient safe behavior at home. Still, the behavior of humans is in great extent defined by environmental and social situations outside of household, depending of the media influence, environmental conditions, political situation, emerging trends of society, etc. All this in various aggregations is influencing the daily human behavior. Therefore, the reasoning abilities of the local (in-house) processing gateways or nodes remain always limited due to lack of holistic information, i.e., the lack of the "big picture." The higher level predictions have to be done in cloud environment which does possess necessary information, reflecting all possible ramifications of the general situation back to the gateway nodes.

The computing capacity problem will be relaxed as the processing capabilities of microcontrollers are improving following the Moore's law. The same is valid in case of various systems on chips (SoCs). For example, the high-end programmable logic SoCs from Xilinx Zync-7000 family have internal block memory capacity till 26.5 Mbits (dual-port, programmable, built-in optional error correction) also they carry Dual-Core ARM Cortex-A9 microcomputer IP [44].

In SW implementation of HTM, a neural connection, i.e., synapse is a data record consisting of an address of source neuron along with additional dynamic information, e.g., permanence and activation history. Here, the permanence represents the stages of growth of the synapse [45]. All this information has to be present in working memory to guarantee efficient processing.

As an example, the Numenta NuPic HTM model tool NuPIC [45] processing network consists typically of 2^{16} neurons, which is matching well to the number of neurons in mammalian cortical column at layer 3 [46]. Thus, an average HTM module (approximate equivalent of cortical column layer 3) consists of 2048 (mini)columns, 32 neurons per column, i.e., 2^{16} neurons $\times$ 20 dendrites per neuron $\times$ 40 synapses per dendrite. Altogether, 52,428,800 words are required to record all possible connections (synapses) in a network. When using regular 32-bit word length, where 16 bits are used as source neuron address of the synapse and 16 remaining bits as a permanence and the activity history value of the synapse, then the minimum amount of allocated memory is 209,715,200 bytes.

Modeling more complex neurons with hundreds of dendrites each with hundreds of synapses, the memory requirement can easily exceed a gigabyte barrier. It is still far from the capacity of top neurons of a human cortex—a single pyramidal cell can have approximately 12,000 dendrites and receive around 30,000 excitatory and 1700 inhibitory inputs [47].

In addition, there are values to represent column activation, state of every neuron, threshold values for dendritic segments and columns, etc. Finally, there are parameters helping to control and distribute loads over all minicolumns and neurons in the network. Processing load is not a serious problem because due to sparsity principle only 2% of neurons will become active after the spatial pooling stage, predicting the next state (input).

A block memory of Xilinx ZYNC SoC with size 8KB allows to record and keep connectivity information about at least a single artificial neuron with its many hundreds of connections (assuming each word to represent source address of synaptic connection and permanence value). The top FPGAs of Xilinx Zynq-7000 series contain many hundreds of block memories (755 in "high"-class XC7Z100, 140 in "consumer"-class XC7Z020) which allows HW implementation of regular HTM. Still, the connection resources on FPGA might become an obstacle although interneuron connections can be implemented using Time Division Multiple Access (TDMA) manner using serial lines only. Integrated multi-core ARM microprocessors are capable of processing even larger amounts of neural processing data in external memory, and the drawback is remarkably higher energetic cost.

Still, the NN processing in regular HW/SW technology is power hungry and not suitable for applications where available energy and memory space is limited. Designer of the power limited system has either decreased the available "smartness" of the nodes or inserted learning capable nodes only to locations with low-speed real-time requirements (allowing majority of time to exploit the sleeping mode), leaving power-hungry tasks to higher level gateways or servers with abundance of power.

As an example, a self-driving car has to process huge amount of high-resolution mapping, visual, radar, and sensory real-time data within tiny fraction of second (faster than a human driver) to guarantee safe driving in unpredictable traffic conditions. In contrast, the processes inside artificial or natural living environments are safe when sampling and decision-making intervals have period in seconds or even minutes. In such circumstances, the usage of sleeping mode allows drastically reduce the overall power consumption (peak power demand has to be satisfied, of course). In analogy with mammals the processing speed of self-driving car controller could be comparable with a cheetah's visual information processing speed, and processes in living environment are advancing in sloth speed. In insect world taxonomy, the extremes might be the fly and caterpillar.

The typical 8- and 16-bit microcontrollers used at the edge of the system (in sensory nodes) are capable only for sensory signal filtering, rudimentary fixed algorithm processing, and communication with higher level nodes due to very limited memory, processing, and power resources. Still, quite complex local regulatory functions are possible at edge nodes, e.g., various PID controllers fall into this category.

For example, the highly popular 32-bit ARM Cortex-M (microcontroller processor) family devices are well suitable for home automation tasks due to integrated peripherals and power efficiency. The only problem is a limited integrated memory (up to 256KB SRAM in TM4C129x MCU series [http://www.ti.com/product/TM4C123GE6PM]), which restricts their usability in neural or cortical learning type of algorithm based applications where the number of tightly interconnected neurons alone might be measured in thousands.

Often more than a single input signal has to be followed to discover an abnormal behavior of the subject. For example, in case of rehabilitation care, the IMUs are used to assess body part movement. The movement might be wrong in any of six degrees of freedom, also along the time dimension (too slow or too fast), which is difficult to represent as an algorithm. Instead of those six separate HTMs, implementing six cortical columns might be able to follow all signals concurrently and the decision-making can be based on aggregated anomaly score of those HTMs.

The HTM is not ready to exhibit full capabilities of the mammalian cortical column. Still, it has a useful set of properties to be exploited at fog-level gateways. The HTM enriched gateways can be responsible for input abstraction, classification, anomaly detection, and communication with cloud level processing nodes, describing the observable situation in high level abstract terms like "normality," "anomaly," "alertness," "attention," "health," etc.

5.9 Conclusion

In this chapter, we have confirmed the observation of many authors that hosting signal analysis and some intelligence in the sensor nodes (edge/mist computing) and the local gateway (fog computing) is beneficial in a variety of applications with respect to performance, reliability, and privacy. The emerging concepts of self-awareness promise to bring a new level of sensible and adaptive behavior to the local sensor and gateways. In all four case studies, these potential benefits of local intelligence are apparent even if they have not been fully realized. However, although the qualitative arguments for distributing computation among the hierarchy levels are compelling, serious challenges remain:

- The involved trade-offs are poorly understood in quantitative terms. Moving a piece of computation from the cloud to the smart gateway or to the sensor node involves a significant change in the communication needs but also in the computation efficiency, because computing platforms are radically different in the different locations. A holistic trade-off analysis will depend on the details of the application, the involved platforms, and the protocols. It has rarely been done for specific cases and is altogether missing in a generally applicable way.
- Although we have many examples of sensor node and gateway architectures, there is no common view on what resources are required and how the architecture should be organized. Some platforms have become fairly popular, but it seems

that the field is moving quickly and requirements are shifting. Thus, convergence on one or two winning platforms is not imminent.
- No methods with supporting tools have been proposed that can guide a an application engineer through the design of application while exploring the trade-offs due to the choice of platforms, functionality, and the distribution of computation across the hierarchy of mist, fog, and cloud computers.

In this chapter, we have illustrated the benefits of intelligence and self-awareness in mist and fog computing and we have sketched a possible architecture for a gateway that meets the requirements as we understand them today. However, this also highlights that there are significant challenges and work to be done in this still young but very dynamic emerging field.

References

1. A.V. Dastjerdi, R. Buyya, Fog computing: helping the internet of things realize its potential. Computer **49**, 112–116 (2016)
2. J.S. Preden, K. Tammemäe, A. Jantsch, M. Leier, A. Riid, E. Calis, The benefits of self-awareness and attention in fog and mist computing. Computer **48**(7), 37–45 (2015)
3. More Data, Less Energy: Making Network Standby More Efficient in Billions of Connected Devices (2014). https://www.iea.org/publications/freepublications/publication/more-data-less-energy.html
4. A. Jantsch, K. Tammemäe, A framework of awareness for artificial subjects, in *Proceedings of the 2014 International Conference on Hardware/Software Codesign and System Synthesis*. CODES '14 (ACM, New York, 2014), pp. 20:1–20:3.
5. N. TaheriNejad, A. Jantsch, D. Pollreisz, Comprehensive observation and its role in self-awareness - an emotion recognition system example, in *Proceedings of the Federated Conference on Computer Science and Information Systems*, Gdansk (2016).
6. IEEE International Conference on Self-Adaptive and Self-Organizing Systems (2007–2016)
7. J. Pitt (ed.), *The Computer after Me: Awareness and Self-Awareness in Autonomic Systems* (Imperial College Press, London, 2014)
8. SelPhyS: workshop on self-aware cyber-physical systems, CPS Week, Vienna (2016)
9. N. Capodieci, E. Hart, G. Cabri, Designing self-aware adaptive systems: from autonomic computing to cognitive immune networks, in *IEEE 7th International Conference on Self-Adaptation and Self-Organizing Systems Workshops (SASOW), 2013* (2013), pp. 59–64
10. S. Kounev, X. Zhu, J.O. Kephart, M. Kwiatkowska, Model-driven algorithms and architectures for self-aware computing systems (Dagstuhl Seminar 15041). Dagstuhl Rep. **5**(1), 164–196 (2015). [Online]. Available: http://drops.dagstuhl.de/opus/volltexte/2015/5038
11. B. Broome, Data-to-Decisions: a transdisciplinary approach to decision support efforts at ARL, in *Proceedings of the Ground/Air Multisensor Interoperability, Integration, and Networking for Persistent ISR III*, vol. 8389 (SPIE, 2012)
12. J.O. Kephart, D.M. Chess, The vision of autonomic computing. Computer **36**(1), 41–50 (2003)
13. A.Y. Zomaya (ed.), *Handbook of Nature-Inspired and Innovative Computing* (Springer, Berlin, 2006)
14. C. Müller-Schloer, H. Schmeck, T. Ungerer (eds.), *Organic Computing — A Paradigm Shift for Complex Systems* (Birkhauser, Basel, 2011)
15. M.T. Higuera-Toledano, U. Brinkschulte, A. Rettberg (eds.), *Self-Organization in Embedded Real-Time Systems* (Springer, Basel, 2013)

16. B. Cheng, R. de Lemos, P. Inverardi, J. Magee (eds.), *Software Engineering for Self-Adaptive Systems*. Programming and Software Engineering (Springer, New York, 2009)
17. D. Vernon, G. Metta, G. Sandini, A survey of artificial cognitive systems: Implications for the autonomous development of mental capabilities in computational agents. IEEE Trans. Evol. Comput. **11**(2), 151–180 (2007)
18. P.R. Lewis, M. Platzner, B. Rinner, J. Torresen, X. Yao (eds.), *Self-Aware Computing Systems: An Engineering Approach* (Springer, New York, 2016)
19. N. Dutt, A. Jantsch, S. Sarma, Towards smart embedded systems: a self-aware system-on-chip perspective. ACM Trans. Embed. Comput. Syst. (2016). Invited. Special Issue on Innovative Design Methods for Smart Embedded Systems
20. N. Dutt, A. Jantsch, S. Sarma, Self-Aware Cyber-Physical Systems-on-Chip, in *Proceedings of the International Conference for Computer Aided Design*, Austin, TX (2015). Invited
21. Vital signs monitoring devices market: Increasing usage in home care settings and sports industry fuelling demand: Global industry analysis and opportunity assessment 2015–2025, London (2015). [Online]. Available: http://www.futuremarketinsights.com/reports/vital-signs-monitoring-devices-market
22. R. Morgan, F. Williams, M. Wright, An early warning scoring system for detecting developing critical illness. Clin. Intensive Care **8**(2), 100 (1997)
23. J. McGaughey, F. Alderdice, R. Fowler, A. Kapila, A. Mayhew, M. Moutray, Outreach and early warning systems (EWS) for the prevention of intensive care admission and death of critically ill adult patients on general hospital wards. Cochrane Database Syst. Rev. **18**(3) (2007)
24. D. Georgaka, M. Mparmparousi, M. Vitos, Early warning systems. Hosp. Chron. **7**(1), 37–43 (2012)
25. A. Anzanpour, I. Azimi, M. Götzinger, A.M. Rahmani, N. TaheriNejad, P. Liljeberg, A. Jantsch, N. Dutt, Self-awareness in remote health monitoring systems using wearable electronics, in *Proceedings of Design and Test Europe Conference (DATE)*, Lausanne (2017)
26. A. Anzanpour, A.M. Rahmani, P. Liljeberg, H. Tenhunen, Context-aware early warning system for in-home healthcare using internet-of-things, in *Proceedings of the International Conference on IoT Technologies for HealthCare (HealthyIoT'15)*. Lecture Notes of the Institute for Computer Science (Springer, Berlin, 2015)
27. M. Götzinger, N. Taherinejad, A.M. Rahmani, P. Liljeberg, A. Jantsch, H. Tenhunen, Enhancing the early warning score system using data confidence, in *Proceedings of the 6th International Conference on Wireless Mobile Communication and Healthcare (MobiHealth)*, Milano (2016)
28. Important Facts about Falls (2016). [Online]. Available: http://www.cdc.gov/homeandrecreationalsafety/falls/adultfalls.html
29. I.-M. Lee et al., Effect of physical inactivity on major non-communicable diseases worldwide: an analysis of burden of disease and life expectancy. Lancet **380**(9838), 219–229 (2012)
30. G. Sprint, D.J. Cook, Using smart homes to detect and analyze health events. Computer **49**, 29–37 (2016)
31. R. Igual, C. Medrano, I. Plaza, Challenges, issues and trends in fall detection systems. BioMed. Eng. Online **12**(1), 66 (2013)
32. L. Dasenbrock, A. Heinks, M. Schwenk, J. Bauer, Technology-based measurements for screening, monitoring and preventing frailty. Zeitschrift für Gerontologie und Geriatrie **49**(7), 581–595 (2016)
33. A. Tessier, M.-D. Beaulieu, C. Mcginn, R. Latulippe, Effectiveness of reablement: a systematic review. Healthcare Policy **11**(4), 49–59 (2016)
34. S. Billinger, R. Arena, J. Bernhardt et al., Physical activity and exercise recommendations for stroke survivors. Stroke **45**(8), 2532–2553 (2014)
35. M.S. Kuster, Exercise recommendations after total joint replacement. Sports Med. **32**(7), 433–445 (2002)

36. G.D. Abowd, A.K. Dey, P.J. Brown, N. Davies, M. Smith, P. Steggles, Towards a better understanding of context and context-awareness, in *Handheld and Ubiquitous Computing (HUC)*, ed. by H.W. Gellersen. Lecture Notes in Computer Science, vol. 1707 (Springer, Berlin/Heidelberg, 1999)
37. J.-P. Vasseur, A. Dunkels, *Interconnecting Smart Objects with IP: The Next Internet* (Morgan Kaufmann, Amsterdam, 2010)
38. P. Harrington, *Machine Learning in Action* (Manning Publications, Greenwich, 2012)
39. S. Astapov, A. Riid, A hierarchical algorithm for moving vehicle identification based on acoustic noise analysis, in *Proceedings of the 19th International Conference Mixed Design of Integrated Circuits and Systems: 19th International Conference Mixed Design of Integrated Circuits and Systems MIXDES 2012* (2012), pp. 467–472
40. D. Graupe, *Deep Learning Neural Networks. Design and Case Studies* (World Scientific, Singapore, 2016)
41. A. Lavin, S. Ahmad, Evaluating real-time anomaly detection algorithms - the numenta anomaly benchmark, in *14th International Conference on Machine Learning and Applications (IEEE ICMLA)* (2015)
42. M.A. Hassan, M. Xiao, Q. Wei, S. Chen, Help your mobile applications with fog computing, in *12th Annual IEEE International Conference on Sensing, Communication, and Networking - Workshops (SECON Workshops)* (2015)
43. pettitda, Road Testing the Raspberry Pi 3 with HTM: Building the Software for 32-bit ARM (2016). https://www.element14.com/community/groups/roadtest/blog/2016/06/07/road-testing-the-raspberry-pi-3-with-nupic
44. Expanding the All Programmable SoC Portfolio (2016). [Online]. Available: https://www.xilinx.com/products/silicon-devices/soc.html
45. J. Hawkins, S. Ahmad, Why neurons have thousands of synapses, a theory of sequence memory in neocortex. Front. Neural Circuits **10**(23), 1–13 (2015). https://doi.org/10.3389/fncir.2016.00023
46. V.B. Mountcastle, The columnar organization of the neocortex. Brain **120**, 701–722 (1997)
47. M. Megías, Z. Emri, T. Freund, A. Gulyás, Total number and distribution of inhibitory and excitatory synapses on hippocampal CA1 pyramidal cells. Neuroscience **102**, 527–540 (2001)

Chapter 6
Urban IoT Edge Analytics

Aakanksha Chowdhery, Marco Levorato, Igor Burago, and Sabur Baidya

6.1 Introduction

The application of the Internet of Things (IoT) paradigm to the urban environment is of particular relevance as it responds to important societal needs and trends [1]. The push to provide solutions toward a functional and efficient Smart City architecture is demonstrated by the large number of academic and industrial endeavors, as well as initiatives from city administrations. For instance, IBM, Siemens, Cisco, ABB, Alcatel-Lucent, Toshiba, and Google have undergoing projects that aim at the development of smart interconnected systems, as well as established city-wide endeavors involving cities in the USA, Europe, and Asia [2, 3].

Current IoT architectures rely on two extremes. On the one hand, enterprise computing largely relies on hauling all the data to the cloud to leverage the cost-benefits and efficiency of a high-capacity storage and compute platform in the data centers [4, 5]. On the other hand, mission-critical applications, such as self-driving cars and autonomous robots, largely rely on local computation for their decision-making because of stringent low latency requirements. In the urban IoT and Smart City scenarios, a city-wide deployment of IoT technologies poses several inherent conceptual and technical challenges that are not resolved by those two extreme architectures. For instance, the transportation of raw streams of data from personal mobile sensors, video surveillance systems, traffic monitoring systems, and other relevant systems to city-scale data centers would require an enormous amount of bandwidth, and energy drain from mobile devices, and would likely result in service

A. Chowdhery
Princeton University, Princeton, NJ, USA
e-mail: aakanksha@princeton.edu

M. Levorato (✉) • I. Burago • S. Baidya
University of California, Irvine, CA, USA
e-mail: levorato@uci.edu

A.M. Rahmani et al. (eds.), *Fog Computing in the Internet of Things*,
DOI 10.1007/978-3-319-57639-8_6

disruption at the wireless edge of the network. Similarly, in a full-scale and mature urban IoT scenario, centralized real-time processing of this large and heterogeneous set of data streams is not feasible.

Edge computing is an architecture that uses end-user clients and one or more near-user edge devices collaboratively to store substantial amounts of data, process compute-intensive tasks, communicate jointly to reduce interference, and carry out management tasks cooperatively to improve application performance. In such edge computing architectures, any device with compute, storage, and networking capabilities can serve as a near-user edge device. The end-user clients and various edge devices can exist in a hierarchy alongside the existing cloud-based architecture to improve the overall system performance. This notion of edge computing is also referred to as "Edge analytics" in [6] or "Fog computing" in [7]. Edge computing solves four key challenges by executing tasks near the edge of local access networks. First, it pools underutilized resources of edge devices in terms of storage or compute capabilities and minimizes the network overhead of hauling data to the cloud. Second, it provides context awareness as application level details are available near the client at the network edge. Thirdly, it enables real-time response with latency in the order of tens of milliseconds by processing near the network edge instead of relying on the cloud, where the multi-hop architecture of the network core may result in undesirable delays. Finally, the software stacks on edge devices can be upgraded in an agile manner without modifying the software stacks in the cloud or core network.

Innovative city-wide architectures should make the best use of those new paradigms, which have the potential to lead to a significant advancement in how data are acquired, transported, and processed over large-scale systems. In particular, there should be a strong interconnection between information acquisition, data communication, and processing across the many geographical and system scales involved. This interconnection can dramatically reduce network load, while significantly improve the quality of Smart City services and reduce response latency. For instance, data fusion and processing performed within or at the edge of local wireless networks can inform data filtering and resource allocation strategies (Fig. 6.1).

The rest of the chapter is organized as follows. Section 6.2 further discusses the design challenges. Section 6.3 presents the architecture and discusses its main components, namely information acquisition and compression, content-aware networking, and information availability. Section 6.5 concludes the chapter.

6.2 Design Challenges

Large cities face many challenges, including traffic congestion, public safety concerns, high energy use, sanitation, public internet connectivity, and providing baseline municipal services. A major issue in establishing smart cities is availability of ubiquitous broadband bandwidth and connectivity. While most modern cities

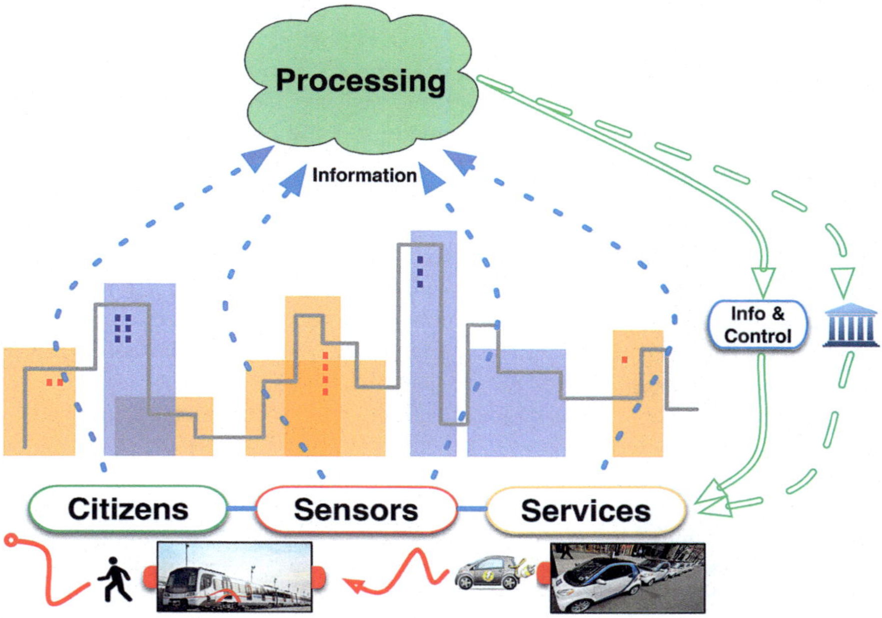

Fig. 6.1 The urban IoT interconnects systems and citizens to provide innovative services

have one or more cellular networks providing adequate coverage, these networks are often designed to have capacity and peak bandwidth limits that just meet the needs of their existing subscribers. This leaves a relatively small, and time-varying, amount of bandwidth for the advanced municipal services envisioned in the Smart City paradigm.

A critical need in modern cities, which is also the focus of Smart city efforts, is safety and security. In the Smart City, this need is addressed by a large and distributed network of sensors and systems. Municipal networks may carry sensitive data (i.e., police dispatches) and operate life-critical systems (e.g., smart transportation, collision avoidance applications, first responder communications, *etc.*), and therefore must be both secure and reliable. Traffic monitoring applications require constant traffic flow updates at each road and intersection to manage road congestion and diverting traffic flows from accident areas.

Related to safety and security, illustrative of the technical challenges of building city-scale systems is video monitoring and surveillance. Smart cities, retail stores, public transport, and enterprises increasingly rely on camera sensors to improve safety and security, identify unauthorized access, and increase reliability of their infrastructure. Local processing does not lend itself to successful deployment, whereas the sheer bandwidth of data being collected over a large-scale network makes it impractical to transport all the data to the cloud to obtain real-time insights. City-scale deployments (such as on traffic lights) and remote areas do not have enough bandwidth to upload high-data rate video streams. Many applications such as real-time tracking and detection of intruders pose strict latency constraints on such infrastructure. Additionally, privacy constraints must be maintained so that the video does not reveal a person's identity to any unauthorized parties. Advanced distributed analytics provides the opportunity to build real-time, latency-sensitive distributed surveillance systems that maintains privacy. We can leverage nearby nodes to intelligently partition video processing between edge devices colocated with cameras and the cloud so as to enable real-time tracking, anomaly detection, and interesting insights from data collected over long time intervals.

Finally, a smart city can also capitalize on the crowdsourced information collected from its citizens using their mobile sensors. For instance, crowdsourced information can be used to estimate parking availability, neighborhood security, wireless signal strength, and congestion in public spaces.

We summarize the key challenges in building a city-wide infrastructure that can capitalize on large sensor systems installed in the city as well as the crowdsourced sensors: *(a) scarce wireless bandwidth*—Available wireless bandwidth is scarce for multiple sensors to coexist with existing wireless services while a wired infrastructure requires heavy investments, *(b) low latency*—Low response latency is critical to applications such as traffic monitoring, where hauling all the data to the cloud to obtain insights can take several minutes to an hour, *(c) efficiency*—A city would require petabytes of storage if it were to transport all the sensor data streams to a single data center, where most of the data would not be useful except to obtain summaries or detect abnormal events. Energy efficiency also favors local computation because the radio transmit power required to continuously communicate collected data often drains the sensors and mobile devices in crowdsourced environments, and *(d) privacy*—Local storage and computation can maintain privacy of the individual sensor streams collected from different entities compared to a centralized solution that aggregates data in a single place. Finally, it is much easier to maintain context of the information closer to the sensor than in a centralized place.

6.3 Edge-Assisted Architecture

Based on the challenges described in the previous section, we contend that edge computing is a key component of such architecture, as it interconnects information acquisition, communication, and computation systems to create a flexible multiscale architecture, where all the components interoperate to maximize efficiency in terms of trade-off between latency, network utilization, energy constraints, and system performance. Intelligence, then, can permeate all the scales of the communication and computation system to enable flexible and adaptive operations targeted to city-wide tasks. The main features of the architecture are:

- *Computation-Aware Information Acquisition*—Edge computing will be the main engine of a distributed intelligent system for adaptive information acquisition. The objective is to preselect and compress data sources across sensor systems to minimize network load and energy expense. The key is to develop algorithms capable of locally removing information which is not needed to accomplish the global computational objective. Note that a similar rationale is used in current algorithms for the compression of multimedia streams, where "information" not usable by humans is removed. In the Smart City context, information irrelevant to the application algorithms can be eliminated. However, different from the former case, in the latter case the needed information is dependent on time-varying parameters such as the computational objective, and the state of the observed system.
- *Content- and Processing-Aware Networking*—In Smart City systems, the objective is to maximize the rate of usable data delivered to the computational resources performing the processing. Edge devices connected to local base stations and access points will enable the implementation of content and processing-aware resource allocation techniques and interference control mechanisms. Interference control can take the form of transmission power and rate control, or channel access control. In both cases, the network manager must be aware of the needs of the application algorithms.
- *Effective information availability*—Edge computing will place data at the edge of the network, thus improving local availability and searchability over a distributed and heterogeneous infrastructure. To this aim, portable semantic structures need to be maintained at all the scales of the system.

In this section, we present the components of the architecture and discuss our preliminary results.

6.3.1 Information Acquisition and Compression

The low-latency link between the edge resources and local sensorial systems enables system-wide messaging determined at the local network scale. The proposed edge-assisted architecture uses this messaging to empower the Urban IoT system with the ability to intelligently and adaptively select relevant data. We contend that the cooperation of devices at different scales is critical to achieve this objective. In fact, whereas IoT devices have all their individual data available, the edge processor may have compressed and corrupted versions of data from a multitude of individual sensors. Thus, the architecture needs to implement messaging to provide contextual information to individual sensors, which will evaluate the relevance of their local data and determine transmission decisions and data compression rate. The data selection and compression system can be seen as a distributed processing system, where heterogeneous agents cooperate to provide critical information to the final application. We logically divide this part of the architecture into adaptive compression and distributed computing components.

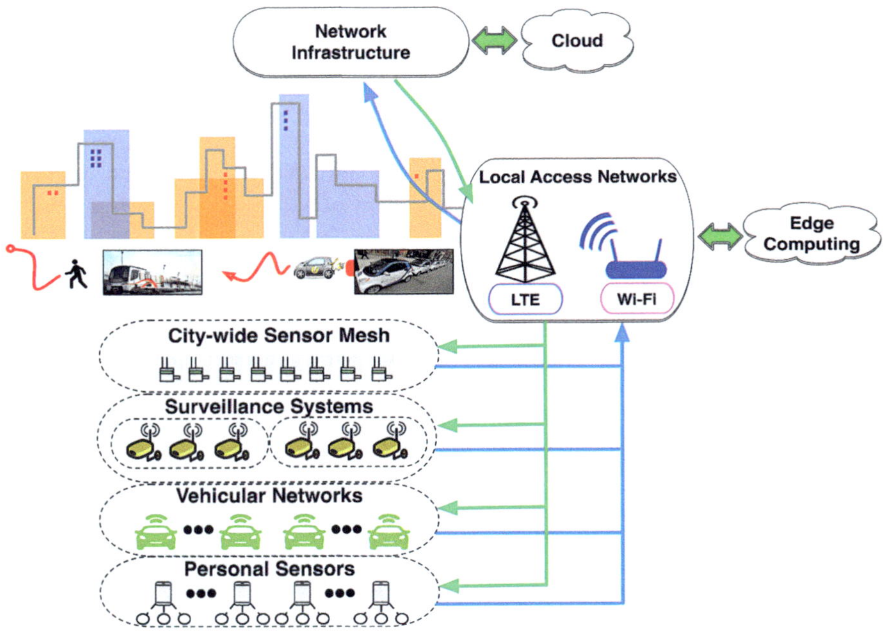

Fig. 6.2 The edge-assisted architecture interconnects information acquisition, communication infrastructures, and processing resources at multiple scales. *Blue* and *green arrows* represent the bidirectional exchange of data and control

Adaptive Compression: One of the key observations to make is that in a sensor-rich, and large-scale, environment where applications have specific computational tasks (e.g., detecting an event), not all the information is needed at the final controller. Conversely, in order to minimize network load and energy expense in the mobile and low-power devices, only necessary information should be pushed through the communication infrastructure (Fig. 6.2).

However, such selection and compression is challenging, due to the scale and heterogeneity of the system, which induces a mismatch between the information available to the individual sensors and the final controller. Being close to the network edge, edge computing can bridge these two extreme scales and support efficient and informed local data selection and compression.

Adaptive compression compresses the sensor streams based on their relevance or importance to the application goal. The incoming sensor stream data is filtered and compressed adaptively. The sensors themselves might lack the compute capability or the storage for prior trained models. However, they extract useful features from sensor streams and communicate them to an edge device. The edge device uses prior trained models to analyze the relevance or priority of the content and signals it back to the sensors for adaptive compression.

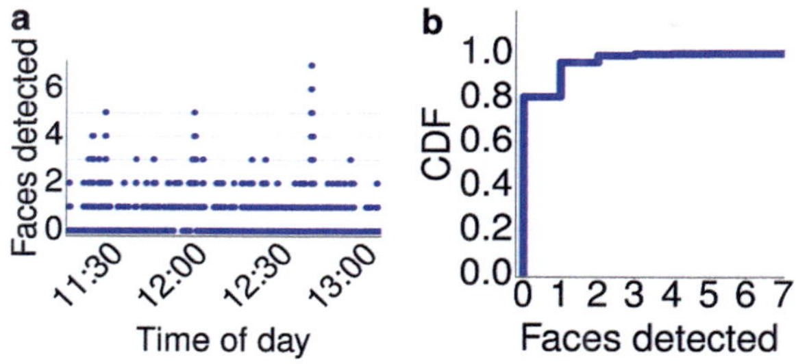

Fig. 6.3 Time series and cumulative distribution function (CDF) of a count of the number of faces a state-of-the-art vision algorithm detects during a busy period at an office building. (**a**) Face count time series. (**b**) Face count CDF

A system, called Vigil [8], illustrates this concept for video monitoring and real-time video surveillance applications. The application goal is to detect the times and detected faces in a busy office hall while the network bandwidth is constrained from the camera device to the central processor. Figure 6.3a shows the number of faces detected at the peak lunch hour, while Fig. 6.3b shows that less than 20% of the collected video in a busy office hall contains relevant information (e.g., moving objects), thus rendering its transportation to the central processor useless. This system uses this insight to implement adaptive compression where the edge device in conjunction with the camera nodes prioritizes the frames with detected faces while sending them over the network.

Note that adaptive compression schemes provide additional gains over MPEG-4 video compression schemes because they are content aware. While standard MPEG-4 video compressions work well for streaming to the web or television, they are not effective for application goals requiring computer vision because artifacts due to spatial and temporal compression impair the effectiveness of algorithms (e.g., object detection, classification, and tracking). Importantly, the activation of the sensor could depend on the state of the observed environment, where more information may be needed if contextual information changes (e.g., a partially hidden moving object is detected). Furthermore, edge computing can be used to interconnect heterogeneous sensor systems, so that bandwidth-demanding sensors (e.g., video capture) are activated based on information from low-bandwidth sensors (e.g., acoustic and motion sensors).

Additionally, the design of adaptive data representation and compression schemes is needed to make them more robust and resilient to the impairments of the wireless channel. This design should be driven by the final objective of processing, where more relevant features, determined based on contextual and local information, are more protected.

Distributed computing among edge devices: The information selection and compression architecture is based on the notion of distributed intelligence. The flexibility granted by edge computing architectures can play an important role in the realization of the envisioned architecture.

In edge computing, a network of sensors and edge devices can be leveraged to allocate computing functions based on their capabilities without requiring the individual sensors to share their sensor streams. This can be extremely valuable

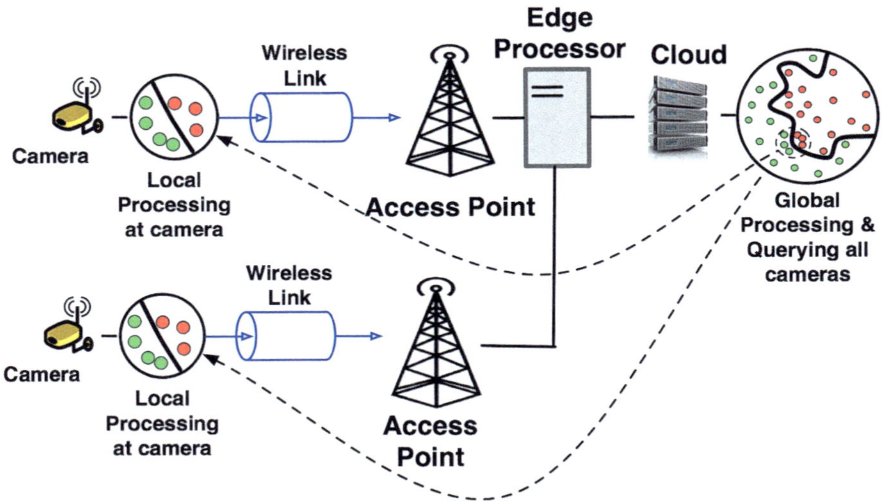

Fig. 6.4 Multi-camera system with distributed classifiers for context-aware local data filtering or selection

in scenarios where the local context can be better established from a group of edge devices and where the storage can be offloaded to a nearby device. The key advantage is that it eliminates the need of using the backhaul network to collect the state information.

At the same time, it brings two new aspects that often come up in distributed computing: *(a)* synchronization of different sensor streams to fuse or correlate them together when collected from devices with different clocks, and *(b)* the granularity of temporal scales to share the information between the edge devices. In the video monitoring example, the video streams largely need to be matched only with respect to detected events or objects when the goal is identifying anomalies or drawing useful insights.

Local computational resources can effectively interconnect individual sensors (e.g., cameras in multi-camera systems), enabling the pruning of these high-bandwidth data streams when only a subset of them is sufficient to perform the tasks dictated by the city applications. Vigil [8], a real-time distributed wireless surveillance, deploys distributed computing across cameras in addition to adaptive compression to improve the system performance. Figure 6.4 illustrates the general architecture. Vigil runs an intra-cluster algorithm across different cameras overlooking the same geographical area to determine the most valuable frames from cameras within a cluster and to eliminate redundant observations, capturing the same objects, to minimize communication bandwidth without actually exchanging the redundant frames. Figure 6.5 illustrates the gains of such an approach. Note that the bandwidth required at low activity level (at most 16 Kbit/s) is lower than the available per-camera wireless capacity and therefore, both Vigil and *Round-Robin* achieve more

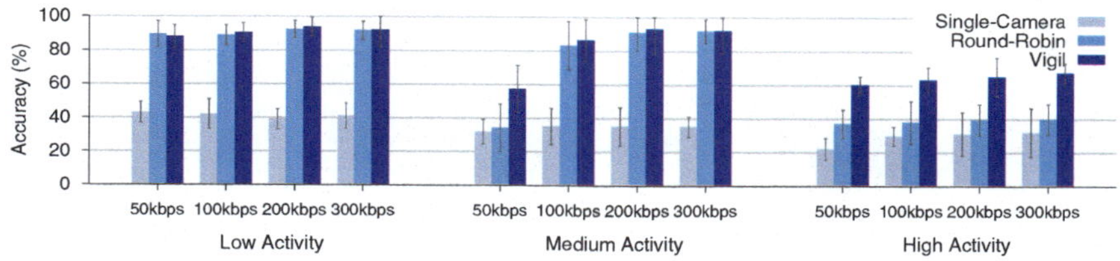

Fig. 6.5 Accuracy of intra-cluster frame selection in system name relative to a single-camera system and a multi-camera system with round-robin scheduling. Error bars show standard deviation of the experiment in varying wireless conditions

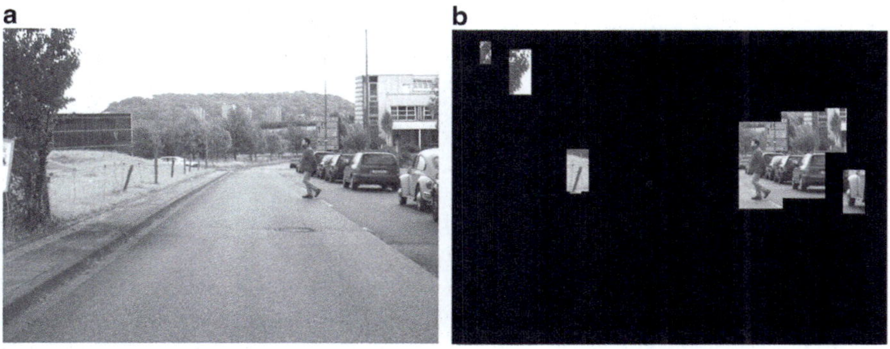

Fig. 6.6 Example of a frame filtered with a Haar feature-based pedestrian classifier. (**a**) Input frame. (**b**) Filtered frame

than 90% accuracy, while the single camera suffers because of lack of sufficient spatial coverage. Similar results are observed for medium activity level, except Vigil outperforms other approaches when the available per-camera wireless capacity 50 kbps is lower than the bandwidth required for medium activity level (at most 80 kbps). Finally at high activity level, the bandwidth required is much higher than the available per-camera wireless capacity and we observe 23–30% gains for Vigil compared to *Round-Robin* because Vigil prioritizes those frames across cameras that maximize the application accuracy.

Extending this reasoning, we studied the energy-bandwidth trade-off in an edge-assisted system, where the video acquisition device is capable of running a simple classification algorithm to eliminate redundant information within each frame. The objective, then, is to transmit to the edge processor only regions within the frames necessary to the global data analysis goal. In the considered setup, the device implements a cascade classifier [9] to select portions of individual frames containing objects of interest (e.g., pedestrians).

Figure 6.6 shows an example of frame before and after the classifier is applied. It can be seen that the classifier mistakenly locates pedestrians in portions containing other objects. This is due to the need to keep the classifier as simple as possible to run on devices with limited computational power while preserving the frame

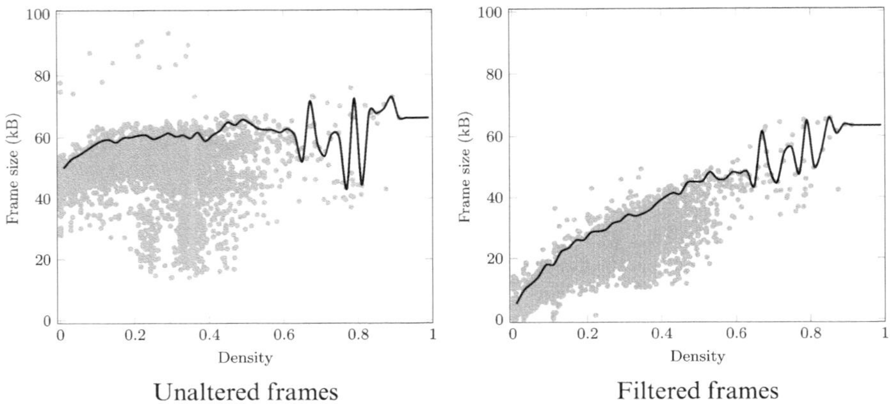

Fig. 6.7 Size of unaltered (**a**) and filtered (**b**) frames in compressed video stream as a function of their object density. *Black lines* depict the corresponding 95% quantiles

output rate. The false positives increase the bitrate requirements when the classifier is activated but do not harm the performance of the remote video processor, which will eventually exclude them by using a more powerful classifier. However, the classifier correctly includes the pedestrian in the picture.

Figure 6.7 depicts the size of the output frames after compression when the classifier is active and inactive. This measure is instrumental to the adaptation framework, as it allows individual sensors to estimate the bandwidth required by the two actions. It can be observed that when the density of objects in the picture is small, the output frame size is much smaller compared to the case when the classifier is inactive. In principle, a perfect classifier would filter out the entire picture when pedestrians are not present. In the practical classifier implemented in our study, when density is smaller than 0.2, the frame size is reduced by a factor of 2 to 10. When the density is large, the activation of the classifier does not correspond to a benefit in terms of output data rate, while increasing energy consumption. The solid lines correspond to the 95% quantile which is used for activation/deactivation decisions.

However, as shown in Fig. 6.8, the video streaming pipeline from the end-device including the classifier requires more power for a longer time with respect to the pipeline without the classifier. There is, thus, an important trade-off between energy spent by edge devices and bandwidth required to stream necessary information to edge and cloud resources. Object density also influences the number of operations necessary to process frames. The energy trade-off is obviously important when mobile devices are considered. However, the overall power consumed by sensors is certainly a central issue in city-scale architectures, where thousands of sensors are deployed.

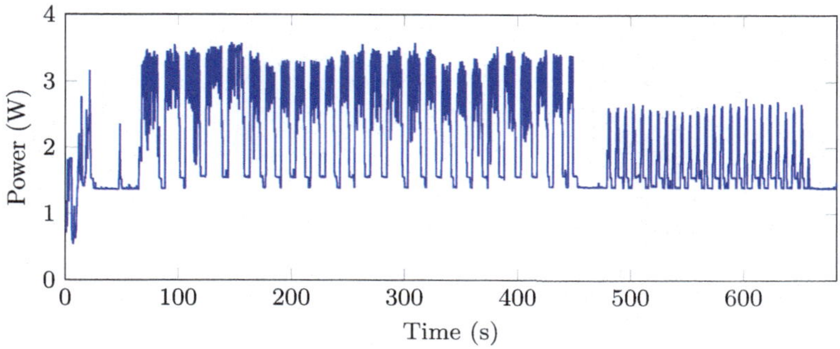

Fig. 6.8 Power consumed by a Raspberry Pi when streaming the video with and without classifier. The clusters of increased power corresponds to group of 60 frames. The first sequence of clusters corresponds to the case where the classifier is used, which increases both energy and execution time

Figure 6.9 depicts exemplar traces for the proposed bandwidth adaptive technique for the two loads. Red lines show the maximum bandwidth available to the end-device in each case, black lines show the actual bandwidth used by the end-device, and gray lines show the bandwidth necessary to transmit unfiltered stream. The strides where the end-device made decision to filter frames are highlighted in green. The effect of bandwidth variations on the output rate are apparent, where a bandwidth insufficient to support the predicted output rate for the current density leads to the activation of the classifier. This action causes visible reduction of the output rates in the plots.

Assistance by the edge processor, then, optimizes the energy-bandwidth trade-off. In the simplest setup, the available bandwidth reported by the local network managers can be used as a constraint, which determines the activation and deactivation of the classifier at the end-device, and how many stages are used. This decision is assisted by the edge, which reports to the end-device the current object density—which can be computed only if the full classifier is used—to enable the prediction of future bandwidth and energy associated with the number of stages performed locally. Note that the local classifier can be re-programmed online to adapt to time-varying application objectives.

6.3.2 Content-Aware Wireless Networking

As discussed in the previous section, an efficient information acquisition architecture is a key component of the urban IoT architecture. However, the transportation of relevant information to the edge computing resource can be a challenging and delicate task due to the complexity and heterogeneity of modern communication infrastructures. Importantly, the urban IoT traffic will share the same network

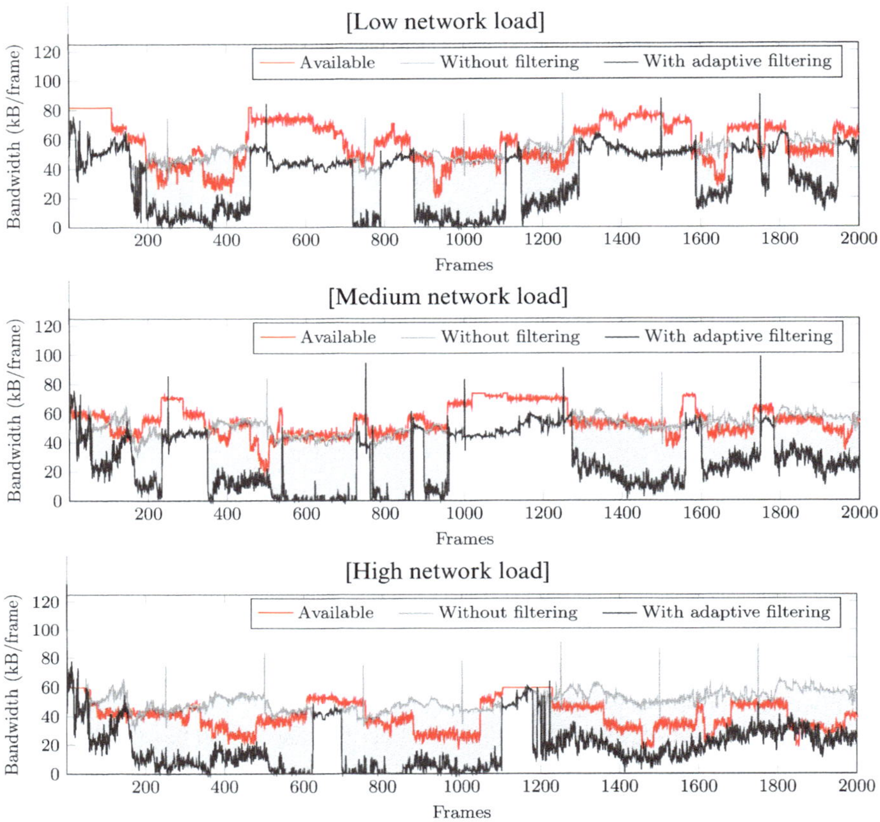

Fig. 6.9 Exemplars of bandwidth traces for streaming adaptively filtered frames in the scenarios of low (**a**), medium (**b**), and high (**c**) network load. *Red lines* show the maximum bandwidth available to the end-device in each case, *black lines* show the actual bandwidth used by the end-device, and *gray lines* show the bandwidth necessary to transmit unfiltered stream. The strides where the end-device decided to filter frames are highlighted in *light gray*

resource with that generated from traditional applications and services. Thus, the communication resources available to the urban IoT may vary over time and be scarce in peak hours. Furthermore, the coexistence of these data streams over a heterogeneous network sharing the same channel resource will make interference control difficult in the physical layer. Hence, we need smart and adaptive network management and aggregation techniques to effectively handle this difficult coexistence.

Mutual interference between information streams can be mitigated by designing access protocols generating specific *content-based* interference patterns. The main idea behind our approach is to make networking and transmission protocols *aware*

Fig. 6.10 In city-scale video processing, artifacts induced by spatial and temporal compression can severely impair the performance of detection and tracking algorithms

of the content being transported and the structural properties of its encoding. This minimizes the loss of relevant information to the processing algorithms given the current state of the observed system.

Our recent work [10, 11] demonstrated the effectiveness of information-centric techniques in coexistence scenarios, where Wi-Fi Device-to-Device (D2D) communications coexist with Frequency Division Duplex (FDD) Long-Term Evolution (LTE) cellular communications on the same bandwidth [12]. The application scenario is city monitoring, where video data streams from surveillance camera systems are processed by real-time edge computational resources to perform object detection and activity recognition [13]. Interference may cause artifacts that would significantly impair the performance of detection and tracking objects (see Fig. 6.10).

Current standards prescribe simple techniques to regulate coexistence in the unlicensed and licensed band. For instance, Wi-Fi and LTE coexistence is realized by implementing listen-before-talk mechanism, where one of the two technologies is prioritized by forcing the other idle when the former is active [14–17]. We contend that more flexible strategies are needed to support the operations of the urban IoT and facilitate coexistence with existing services. For the coexistence in licensed bands, recent work proposes scheduling and interference control strategies that limit the Signal-to-Interference-plus-Noise-Ratio (SINR) at the cellular base stations [18–20]. However, these techniques often require instantaneous channel knowledge and may result in packet loss when coordination between networks is not perfect. Our design revolves around the notion of utility of data within the stream, where utility variables are computed, exchanged, and processed by the cloud, edge, and end-device resources. The colocation of the edge computation resource and network controllers such as base stations and access points allows to establish a direct exchange of information between them. The edge, then, processes and communicates the utility to associated network controller, which determines channel allocation and transmission strategy of the connected end-devices based on the current network state. However, the operations of this class of protocols require information about content being transmitted and processing state. To this aim, content and state information should be shared with network transmitters and resource management units to create content-specific interference patterns and resource allocation.

Conforming with the 3rd Generation Partnership Project (3GPP) standard for proximity services [21], we choose a topology where an end host is transmitting real-time data on the uplink of LTE to the Internet for computation and processing, and two mobile devices in proximity are connected with each other with network-assisted D2D communications. The LTE end-user is streaming real-time video on uplink by Evolved UMTS Terrestrial Radio Access Network (E-UTRAN) toward the edge resource. The LTE base station (element Node B—eNodeB) scheduler allocates resource blocks for data transmission and assigns modulation and power based on channel quality and interference [22]. When the D2D communication interferes with the LTE uplink, the channel quality degrades and the LTE receiver will more likely fail to decode the packet. Note that interference from the D2D link may influence the modulation and transmission power of the User Equipment (UE).

Video compression techniques exploit spatial and temporal similarities in individual frames and across video. In the most efficient compression standard, H.264 creates Group of Pictures (GoP) composed of reference (I-Frame) and differential (P- and B-Frames) frames. Reference frames transport the whole picture, whereas differential frames encode differences with respect to the reference. When an encoded frame is damaged, due to spatial compression, it affects the transform coefficients which leads to the corruption in the decoded image. The spatial propagation of errors may create artifacts that are detected as objects or impair the ability of the algorithm to detect existing objects. If a reference frame is corrupted, the effect propagates through the entire GoP. When a differential frame is damaged, the effect is smaller compared to loosing part of a reference frame, as key features in the following frames may be recovered using the reference frame. In the proposed framework, we use a simple notion of utility based on frame class, where the edge decompresses the video stream prior to processing and signals to the eNodeB when a reference or differential frame begins. Based on this information and channel statistics, the eNodeB determines the transmission probability of the D2D link. Thus, channel access in the local network is regulated based on the transmitted data, and based on the feedback from the computation algorithm consuming the data stream.

Figure 6.11 shows object detection probability as a function of the throughput of the D2D link for fast and slow fading channels, where the data stream transports a video from a parking lot surveillance camera. The content-based transmission probability scheme (Frame Determined Transmission Probability—FDTP) is compared with the case where the D2D transmits with Fixed Probability (FP). In the considered case, video transmission consumes the entire LTE bandwidth, and the D2D link would starve if listen-before-talk is used. The lines are obtained by varying the transmission probabilities. For comparable object detection performance, the

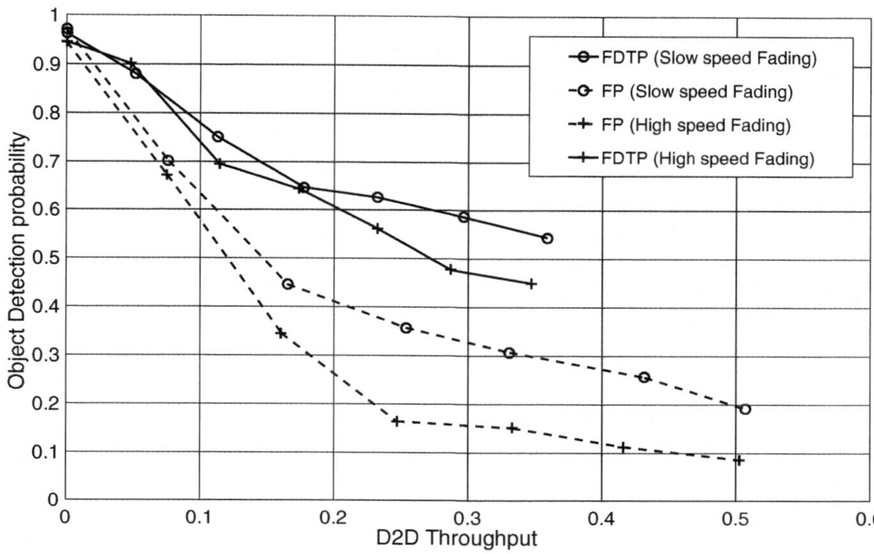

Fig. 6.11 Object detection probability vs D2D throughput in low speed and high speed fading scenario

FDTP scheme grants significant throughput increase to the D2D link, thus enabling coexistence on the same resource. Our study in [10] shows that the efficiency of coexistence, measured as application performance over throughput of the D2D link, is maximum in specific transmission power and probability regions.

6.3.3 *Information Availability*

While the sensors themselves can use the data they collect to make intelligent decisions, the edge devices and the cloud have access to a larger pool of sensor data from multiple sensors and at multiple timescales. Thus, the cloud and the edge devices can assist in making intelligent decisions that individual sensors might be incapable of making. Further, since the cloud has longer range information in temporal scales, it can understand the traffic patterns and other unexpected events such as roadwork or accidents, to plan more efficient paths in traffic monitoring scenarios.

Existing literature relies on aggregating the data in the cloud or a cluster of thousands of servers that can be indexed to enable structured or nonstructured data queries. The cloud-based models are heavily used for applications such as web search, advertising, social networking, and photo repositories to enable users to query data or draw insights. These models largely rely on distributed dataflow systems and programming models, e.g., Map-Reduce [23] and Spark [24]. Sensor data and video data can be especially challenging to search in, for example, existing methods to analyze the video feeds in real time or post facto do not scale and are error-prone. The vision pipelines must be carefully handcrafted for cloud execution by the engineers requiring their focus on nitty gritty details such as how to parallelize, in which order to execute the modules, etc. Similarly, existing dataflow systems such as Spark require analogous handcrafting of user-defined modules as they lack query optimization. Supporting ad hoc queries or post facto analysis on stored video or scaling to a large number of cameras remain key open problems.

A recent research paper proposes a system Optasia [25] that brings together advances from two areas: machine vision and big data analytics systems. This convergence leads to an efficient query answering system over many cameras. The system demonstrates a modularized approach to building vision processing components for applications such as classifying vehicles by color and type, reidentifying vehicles across cameras, tracking lane changes, identifying license plates, etc. This modularized implementation allows the dataflow system to de-duplicate and parallelize the processing.

To address the challenge of scaling to a rich set of ad hoc queries and to many cameras, Optasia casts the problem as an application of a relational parallel dataflow system and wraps the above-described vision modules inside some well-defined interfaces (processors, reducers, and combiners). This makes querying efficient and fast for a city-wide deployment of cameras by decomposing the vision analytic tasks. Each vision module is expressed as a composition of the corresponding relational operators (select, project, aggregate, and Cartesian product). End-users simply declare their queries over the modules in a modified form of SQL. Then, the query optimizer reuses optimization rules and translates user queries into appropriate parallel plans over several different vision modules. The primary advantages of this combination are: (1) ease-of-use for end-users, (2) decoupling of roles between end-users and the vision engineers: the vision engineers can ignore pipeline construction and need only focus on efficiency and accuracy of specific modules, and (3) automatic generation of appropriate cloud execution plans that, among other things, de-duplicate similar work across queries and parallelize appropriately.

Evaluation on traffic videos from a large city on complex vision queries shows high accuracy for Optasia with many fold improvements in query completion time and resource usage relative to existing systems. Figure 6.12a plots the ratio of the query completion time for Optasia with query optimization against a version of Optasia that has no query optimization, for single queries on a parking garage video feed. We see that, with query optimization, Optasia is roughly 3× faster. Further, the query completion time of Optasia remains constant as dataset sizes increase illustrating the fact that the query optimization sets the degree-of-parallelism correctly. The large gains arise from de-duplicating the work in the vision modules (e.g., generating histogram-of-gradient features, etc.).

Within this area, several challenges must be solved to develop systems such that make information readily available at city-scale and can answer queries in real-time. In the multiscale edge architecture we proposed, in addition to analyzing sensor

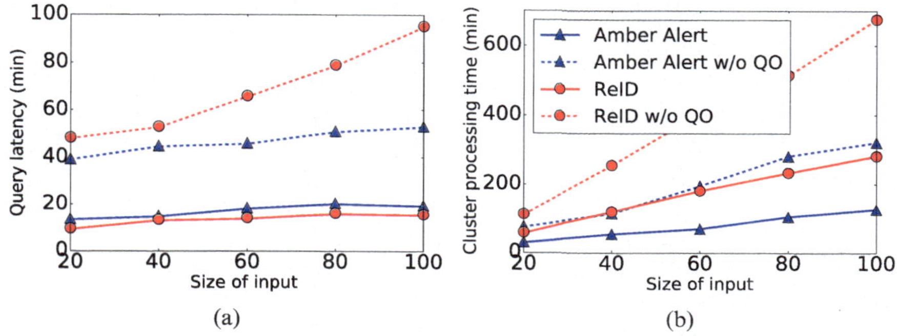

Fig. 6.12 In Optasia [25], Query Optimization reduces the query completion time significantly for both amber alert and Re-ID (**a**) as the number of input videos increases for each query. Further, query optimization ensures the most efficient cluster resource utilization in terms of processing time (**b**)

streams, the key challenge is that we need to search, identify anomalies, trigger alerts, and draw insights from the data collected over a multitude of edge nodes. To enable efficient ways to search over a multitude of distributed sensors, the key questions we must answer are: (a) How do we represent different spatial regions in the environment at the urban scale?, (b) How does the cloud obtain information about regions at different resolutions?, and (c) How do intermittently connected sensors transmit sensor information from these regions in a loss-resilient manner over the wireless channel?. One approach is similar to the compression approach used by Graphics community, where they use "Octree" to compress 3D content, especially point clouds to represent 2D spatial grids. Recent works [26] have shown that the approach extends to collecting and querying the sensory data collected by the self-driving cars. An alternate approach relies on Named Data Networking to name information objects and make them easy to query by other objects.

An additional challenge that arises in drawing insights from a multitude of sensors is to ensure appropriate access control mechanisms and respect data privacy where needed. Thus, computation techniques that are privacy-preserving, such as secure multiparty computation [27] or differential privacy [28], can be extremely useful in data aggregation over a variety of sources.

6.4 Related Work

Several architectures have been proposed which process Smart City data in the cloud [4]. Related to the application case discussed herein, in [29], an adaptive architecture to discover the topology of a distributed multi-camera system is presented. Mitton et al. [5] proposes a general cloud-based architecture for distributed sensor systems in the Smart City.

Edge and fog computing techniques have been considered to effectively reduce the load generated to the central communication and computation infrastructure [6, 7, 30, 31]. However, the solutions explored so far solely focus on data processing, without an in-depth analysis of information acquisition, representation, and transportation solutions needed to increase efficiency and achieve a sustainable technology.

Another line of work [32, 33] aims to reduce the latency of uploading data to the cloud, by partitioning computation tasks between mobile sensors and the cloud for personal mobile devices. Odessa [32] supports interactive perception applications by dynamically offloading parts of computation tasks from mobile devices to the cloud. A recent system Gabriel [33] targets a similar class of augmented reality applications based on a cloudlet architecture, which comprises computation devices located at the edge of network to reduce network latency.

Sensor selection in microscale sensor networks, e.g., see [34], and in-network compression, e.g., [35], have been the focus of intense work. However, we contend that these approaches do not directly apply to the urban IoT scenario. Although these works provide solid theoretical and system design basis, the involved multisystem, multiscale urban IoT architecture requires significant conceptual and practical advancements.

6.5 Conclusions

In this chapter, we proposed a novel architecture supporting urban IoT operations based on our prior results and experimental experience. One of the main contributions is the notion that the information acquisition, networking, and computation logical components of the urban IoT should be interconnected and conjointly operate to make city-wide applications feasible. The proposed architecture, then, is based on a notion of intelligence that pervades all the layers and devices operating in the urban IoT and uses edge computing as a key element to bridge the local fine-time scale of sensors to the coarser topological and temporal operations of the cloud. We introduced the notion of context and computation-aware data selection and compression to maximize the efficiency of the communicated data for specific applications and processing tasks. We introduced the concept of content-aware networking protocols that tune channel access and transmission based on the representation and relevance of the data travelling over the network. Finally, we argued that the presented layered architecture will facilitate information search and improve its availability.

References

1. A. Zanella, N. Bui, A. Castellani, L. Vangelista, M. Zorzi, Internet of things for smart cities. IEEE Internet Things J. **1**(1), 22–32 (2014)
2. P. Neirotti, A.D. Marco, A. Cagliano, G. Mangano, F. Scorrano, Current trends in smart city initiatives: some stylised facts. Cities **38**, 25–36 (2014)
3. M. Naphade, G. Banavar, C. Harrison, J. Paraszczak, R. Morris, Smarter cities and their innovation challenges. Computer **44**(6), 32–39 (2011)
4. M. Rahimi, J. Ren, C. Liu, A. Vasilakos, N. Venkatasubramanian, Mobile cloud computing: a survey, state of art and future directions. Mobile Netw. Appl. **19**(2), 133–143 (2013)
5. N. Mitton, S. Papavassiliou, A. Puliafito, K. Trivedi, Combining cloud and sensors in a smart city environment. EURASIP J. Wirel. Commun. Netw. **2012**(1), 1–10 (2012)
6. M. Satyanarayanan, The emergence of edge computing. Computer **50**(1), 30–39 (2017)
7. Openfog reference architecture for fog computing, produced by the openfog consortium architecture working group. [Online]. Available: https://www.openfogconsortium.org/ra/
8. T. Zhang, A. Chowdhery, V. Bahl, K. Jamieson, S. Banerjee, The design and implementation of a wireless video surveillance system, in *Proceedings of the 21st Annual International Conference on Mobile Computing and Networking* (ACM, New York, 2015), pp. 426–438
9. K.-D. Lee, M.Y. Nam, K.-Y. Chung, Y.-H. Lee, U.-G. Kang, Context and profile based cascade classifier for efficient people detection and safety care system. Multimed. Tools Appl. **63**(1), 27–44 (2013)
10. S. Baidya, M. Levorato, Content-based cognitive interference control for city monitoring applications in the urban IoT. IEEE Globecom 2016, Dec 4–8, Washington, DC, 2016
11. S. Baidya, M. Levorato, Content-based interference management for video transmission in d2d communications underlaying LTE, in *IEEE ICNC 2017*, Jan 26–29, Silicon Valley, 2016
12. K. Doppler, M. Rinne, C. Wijting, C.B. Ribeiro, K. Hugl, Device-to-device communication as an underlay to LTE-advanced networks. IEEE Commun. Mag. **47**(12), 42–49 (2009)
13. S. Hengstler, D. Prashanth, S. Fong, H. Aghajan, Mesheye: a hybrid-resolution smart camera mote for applications in distributed intelligent surveillance, in *Proceedings of the 6th International Conference on Information Processing in Sensor Networks* (ACM, New York, 2007), pp. 360–369
14. J. Jeon, H. Niu, Q. Li, A. Papathanassiou, G. Wu, LTE with listen-before-talk in unlicensed spectrum, in *2015 IEEE International Conference on Communication Workshop (ICCW)* (IEEE, New York, 2015), pp. 2320–2324
15. R. Ratasuk, N. Mangalvedhe, A. Ghosh, LTE in unlicensed spectrum using licensed-assisted access, in *2014 IEEE Globecom Workshops (GC Workshops)* (IEEE, New York, 2014), pp. 746–751
16. A. Mukherjee, J.-F. Cheng, S. Falahati, L. Falconetti, A. Furuskär, B. Godana, H. Koorapaty, D. Larsson, Y. Yang et al., System architecture and coexistence evaluation of licensed-assisted access LTE with IEEE 802.11, in *2015 IEEE International Conference on Communication Workshop (ICCW)* (IEEE, New York, 2015), pp. 2350–2355
17. R. Ratasuk, M.A. Uusitalo, N. Mangalvedhe, A. Sorri, S. Iraji, C. Wijting, A. Ghosh, License-exempt LTE deployment in heterogeneous network, in *2012 International Symposium on Wireless Communication Systems (ISWCS)* (IEEE, New York, 2012), pp. 246–250
18. P. Phunchongharn, E. Hossain, D. Kim, Resource allocation for device-to-device communications underlaying LTE-advanced networks. IEEE Wirel. Commun. **20**(4), 91–100 (2013)
19. C. Yu, O. Tirkkonen, K. Doppler, C. Ribeiro, On the performance of device-to-device underlay communication with simple power control, in *IEEE 69th Vehicular Technology Conference*, pp. 1–5, 2009
20. Y. Wen-Bin, M. Souryal, D. Griffith, LTE uplink performance with interference from in-band device-to-device (D2D) communications, in *IEEE Wireless Communications and Networking Conference*, pp. 669–674, March 2015

21. 3GPP TR 36.843 feasibility study on LTE device to device proximity services - radio aspects (2014)
22. European Telecommunications Standards Institute, E-UTRA physical layer procedures, Generation Partnership Project Technical Specification (3GPP TS) 36.213, V.10, 2011
23. D. Jiang, B. Ooi, L. Shi, S. Wu, The performance of mapreduce: an in-depth study. Proc. VLDB Endow. **3**(1), 472–483 (2010)
24. M. Armbrust et al., Spark SQL: relational data processing in spark. in *SIGMOD* (2015)
25. Y. Lu, A. Chowdhery, S. Kandula, Visflow: a relational platform for efficient large-scale video analytics, in *ACM Symposium on Cloud Computing (SoCC)* (ACM, New York 2016)
26. S. Kumar, L. Shi, N. Ahmed, S. Gil, D. Katabi, D. Rus, Carspeak: a content-centric network for autonomous driving. SIGCOMM Comput. Commun. Rev. **42**(4), 259–270 (2012) [Online]. Available: http://doi.acm.org/10.1145/2377677.2377724
27. C.-T. Chu, J. Jung, Z. Liu, R. Mahajan, sTrack: secure tracking in community surveillance, in *Proceedings of the 22nd ACM International Conference on Multimedia*. MM '14, pp. 837–840, 2014
28. C. Dwork, K. Kenthapadi, F. McSherry, I. Mironov, M. Naor, Our data, ourselves: privacy via distributed noise generation, in *Proceedings of the 24th Annual International Conference on the Theory and Applications of Cryptographic Techniques*. EUROCRYPT'06, 2006
29. Y. Wen, X. Yang, Y. Xu, Cloud-computing-based framework for multi-camera topology inference in smart city sensing system, in *Proceedings of the 2010 ACM Multimedia Workshop on Mobile Cloud Media Computing* (ACM, New York, 2010), pp. 65–70
30. M. Satyanarayanan, P. Simoens, Y. Xiao, P. Pillai, Z. Chen, K. Ha, W. Hu, B. Amos, Edge analytics in the internet of things. IEEE Pervasive Comput. **14**(2), 24–31 (2015)
31. F. Bonomi, R. Milito, J. Zhu, S. Addepalli, Fog computing and its role in the internet of things, in *Proceedings of the First Edition of the MCC Workshop on Mobile Cloud Computing*. MCC '12, pp. 13–16, 2012
32. M.-R. Ra, A. Sheth, L. Mummert, P. Pillai, D. Wetherall, R. Govindan, Odessa: enabling interactive perception applications on mobile devices, in *Proceedings of the 9th International Conference on Mobile Systems, Applications, and Services*. MobiSys '11 (ACM, New York, NY, 2011), pp. 43–56. [Online]. Available: http://doi.acm.org/10.1145/1999995.2000000
33. K. Ha, Z. Chen, W. Hu, W. Richter, P. Pillai, M. Satyanarayanan, Towards wearable cognitive assistance, in *Proceedings of the 12th Annual International Conference on Mobile Systems, Applications, and Services*. MobiSys '14, 2014, pp. 68–81
34. U. Mitra, B. Emken, S. Lee, M. Li, V. Rozgic, G. Thatte, H. Vathsangam, D. Zois, M. Annavaram, S. Narayanan et al., Knowme: a case study in wireless body area sensor network design. IEEE Commun. Mag. **50**(5), 116–125 (2012)
35. G. Quer, R. Masiero, G. Pillonetto, M. Rossi, M. Zorzi, Sensing, compression, and recovery for WSNs: sparse signal modeling and monitoring framework. IEEE Trans. Wirel. Commun. **11**(10), 3447–3461 (2012)

Part IV
Application Use-Cases

Chapter 7
Control-as-a-Service in Cyber-Physical Energy Systems over Fog Computing

Korosh Vatanparvar and Mohammad Abdullah Al Faruque

7.1 Power Grid and Energy Management

Power grid or an electrical grid is an interconnected network for delivering electricity from suppliers to consumers. It consists of generating stations or power plants that produce electrical power, transmission lines carrying power from suppliers to demanding centers, and distribution lines powering individual customers and users in the grid. However, power grid nowadays is not only made of large suppliers and distant customers, it may also consist of various electrical components in different levels (e.g., transmission and distribution). In a distribution grid, there may be a group of distributed electrical energy resources (storage and generation) and loads (devices and appliances) which may operate independently (island mode) or connected to the main grid. This localized distributed energy system is called microgrid [1–3]. The primary purpose of a microgrid is to ensure local, reliable, and affordable energy security for communities. For instance, microgrids have been deployed significantly at residential areas besides military installations, critical infrastructure areas (e.g., hospitals), commercial, and university locations [4, 5]. Consumer EV, rooftop photovoltaic systems, residential-scale energy storage, and smart flexible appliances constitute the driving technologies in forming a residential microgrid [6].

The growing number of devices and customers in the power grid increases the demand for electrical energy significantly. Figure 7.1 illustrates the detail of contributions to energy consumption. For instance, about 41% of the US primary energy consumption is due to commercial buildings and residential homes [7]. About 65% of the electricity in the USA is generated from coal, gas, and other

K. Vatanparvar (✉) • M.A. Al Faruque
Department of Electrical Engineering and Computer Science, University of California, Irvine, CA 92697, USA
e-mail: kvatanpa@uci.edu; alfaruqu@uci.edu

A.M. Rahmani et al. (eds.), *Fog Computing in the Internet of Things*,
DOI 10.1007/978-3-319-57639-8_7

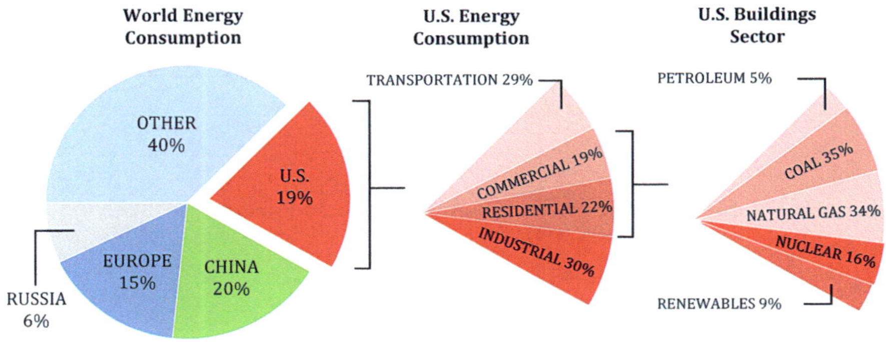

Fig. 7.1 Contribution to energy consumption in the USA

fossil fuels [8]. This will have a huge influence on the environment condition in terms of global warming, air quality deterioration, oil spills, and acid rain [9, 10]. Hence, there is a global push towards reducing the energy consumption overall and pushing towards using renewable energy from various sources (e.g., solar, wind, and geothermal). For instance, all retail suppliers of electricity in California should meet the goal of 33% eligible renewable energy by 2020 [11]. Moreover, EPA is regulating state-specific CO2 emission goal that may increase the non-hydro renewable energy capacity about 50% by 2020 [12]. Moreover, reducing the energy consumption of the buildings and homes will decrease this number significantly. Towards this direction, the California Energy Commission (CEC), planned to reach Zero Net Energy (ZNE) buildings by 2020 [13]. Also, the US Department of Energy (DOE) is developing technologies and techniques to improve the efficiency of the new and existing residential and commercial homes and buildings and thereby reduce the national energy consumption [14]. Hence, the need for energy management in order to make a balance between electricity demand and supply and making energy efficient buildings and homes is growing rapidly.

7.1.1 Energy Management Methodologies

There are various methodologies leveraged by utilities to manage the supply and demand while reducing energy consumption. These methodologies are implemented in different levels of a power grid (e.g., microgrid) and can be categorized as:

- **Smart and energy efficient appliance control:** some of the devices and appliances in the power grid are becoming smart enough such that they can sense their environment and make optimum decision on their control inputs. This device-level control attempts to reduce the energy consumption of the device while providing the service to the users. Some of these controllers are: smart lighting, smart heating/cooling, smart washing, and smart Electric Vehicle (EV)

charging. In these mentioned controllers, the device will operate and consume energy according to how many occupants are available and how much of a demand may be predicted in the future by the user.
- **Utilizing renewable energy:** renewable energy resources are the alternatives which may help the environment while providing the energy required for the users. However, the usage of renewable energy comes with its inherent challenges including intermittency, high cost, unreliability, etc. [15]. Research has been very active in this domain to come up with ideas to address these challenges [15]. Therefore, there are various energy management techniques in this level in order to gain the most renewable energy and efficiency.
- **Electricity demand management:** there is always an imbalance between the power supply and demand in the power grid which is costly and challenging to address for the utilities. This becomes more significant by introduction of distributed energy systems such as renewable energy resources and EV charging stations. Therefore, the utilities benefit from different methodologies to manage the demanded energy by the users. They may utilize Demand Response (DR) or Direct Load Control (DLC) methodologies and send a signal to the customers in order to force them reduce their energy consumption or schedule their consumption to another time when there is more supply. On the other hand, the utilities may use an incentive-based methodology where they offer cheaper electricity rate for the time when there is more supply. This methodology is typically referred to as Time-of-Use (ToU) rate policy which is becoming common in many utilities [16–19].

These energy management methodologies are typically implemented by the utilities or third-party electrical companies by the request from the customers. Most of the time, customers may pay monthly fees and/or installation cost for the services and the incentives are the reduction in their final electricity cost.

7.2 Cyber-Physical Energy Systems

Traditional power grid and energy management methodologies are not capable of handling the new growing, distributed, and imbalanced power demand and supply. Hence, there has been a paradigm shift from the traditional, noninteractive, manually controlled power grid to tight integration of both cyber and physical systems [3, 20, 21]. The cyber system brings the capability of computation, communications, and control—discrete dynamics, while the physical system defines the flow of electricity governed by the laws of physics—continuous dynamics. The smart grid which leverages this cyber and physical integration is called ***Cyber-Physical Energy System (CPES)***. The CPES brings the multilevel monitoring and control capability to the power grid in order to improve its reliability, flexibility, efficiency, and cost-effectiveness [22, 23]. The multilevel energy management is also beneficial for addressing the power grid complexity and challenges introduced by distributed energy resources and dynamic electricity loads [10, 24, 25].

Capabilities of monitoring, control, computation, and connectivity are the preliminary requirements of a platform for implementing these multilevel energy managements for CPES. There might be multitude of sensors at different levels of the system (e.g., device and microgrid) responsible for monitoring multiple variables indicating the state of the system. All the sensed data may be transmitted and stored in controllers. They are responsible for processing the data as part of the energy management. The optimum control inputs as the result of the data process are applied to the system using actuators. The connectivity between the sensors, actuators, and controllers is provided through wired or wireless channels using various protocols (e.g., LAN, WiFi, and ZigBee).

Implementation of energy management in CPES has its own challenges. There are various important properties to be considered for implementing the platform. One of them is the probability of penetrating into the consumer market and affordability of the platform for an ordinary consumer. Major requirements for the architecture which influence this penetration and affordability are: (1) interoperability; (2) scalability; (3) ease of deployment; (4) open architecture; (5) plug-n-play capability, and (6) local and remote monitoring [26, 27]. Moreover, meeting these requirements in a single package should also be cost-effective. This becomes more severe for consumers in residential buildings which have limited budget and space for deploying the platform.

Home Energy Management (HEM) as an example for energy management in a CPES has to meet the above-mentioned policies and rules. Since the market of HEM in residential buildings and homes has one of the largest number of customers, the requirements for the platform become more stringent. Different hardware, software, and communication architectures have been proposed and compared in terms of their power consumption, performance, and other factors [28–33]. However, the implementation cost and effort for such a platform including the computing devices, software stack, and communication devices may be high enough that hinder the process of deploying it for ordinary residential users. Control4, Honeywell, and various other companies [34–37] are providing the customers with HEM platforms which transform an existing home to a smart home. These products implement various functionalities such as: temperature control, efficient lighting, and management of smart devices. All these features are provided by a single device, despite the fact that the hardware and software architectures may not be able to handle the growing number of sensors and actuators alongside their heterogeneity. Hence, the scalability, adaptability, and cost of the platforms may become an issue for customers. For instance, the products limit the number of devices connected for management and processing. As a result, scaling the products for larger homes may become significantly expensive. Furthermore, the users do not have the option of customizing the services they require and they have to purchase the whole package as it is.

In summary, the major challenges with a platform design for implementing an energy management, especially multilevel management, are as follows:

- There are various types of devices in a CPES with different purposes and communication protocols. Hence, bringing interoperability and interactivity among these heterogeneous devices in energy management platform with high performance is challenging.
- The ability to customize the services, adapt the platform for different applications, and scaling it for various types of buildings, homes, and areas are important features of the platform.
- The cost of implementing the energy management platform, hardware, and software stack affects the affordability of the platform by the users which defines the probability of penetration into the market.

7.3 Internet-of-Things and Fog Computing

As discussed in the previous section, a platform for an energy management system should have certain features and meet specific requirements. These are necessary for making the platform economical and affordable for ordinary costumers.

Some of the main requirements for the platform are the connectivity, interoperability, and local/remote monitoring and control. Recently, various communication technologies have provided the needed infrastructure. Moreover, the advancement in technology, and the possibility of integrating low-power and high performance electronic devices have enabled us to build advanced embedded systems. Also, reduction in the cost and size of devices like: sensors, actuators, network adapters, and switches has provided us with the opportunity to build sophisticated and low-cost energy management systems. These factors contribute to making the implementation easier and more economical for wider range of customers. Hence, the connection between intelligent and self-configurable devices in a dynamic and global network platform brings the new paradigm which is introduced as ***Internet of Things (IoT)***. IoT overcomes the gap between the physical world and their representation in information systems by providing the required connectivity and interoperability [38–43].

Hence, IoT has been seen as a possible platform for modern technologies, e.g., energy management system, smart home, and smart grid. The IoT is not specific to CPES and energy management and can be adapted to other areas and applications as well. For instance, the vehicles are also getting equipped with a lot of sensors (e.g., GPS location and road conditions) and communication devices. The devices are utilized to provide dynamic mobile communication systems that communicate between vehicles and public networks using V2V (vehicle-to-vehicle), V2R (vehicle-to-road), V2H (vehicle-to-human), and V2S (vehicle-to-sensor) interactions. As in all IoTs, the interconnectivity and data sharing can effectively guide management systems (e.g., energy management) to monitor, supervise, and efficiently control the devices at different scales and levels without the need for human interference.

IoT may contain billions of devices that can sense, communicate, compute, and actuate. Although, the number of devices is highly dependent on the domain of the IoT operation, e.g., home or microgrid, it is growing rapidly over the years, even within a home. For instance, an HEM monitors and controls the devices within a home [38]; everyday, more smart devices are added to the homes which are capable of getting connected to the internet and providing the capability of being controlled locally or remotely. The connection and addition of devices introduce more flexibility for managing the energy within homes and microgrids.

These capabilities have changed the traditional and manual way of controlling towards more cyber-integrated controlling. However, it needs to be noted that there is huge amount of data produced within the IoT by the sensors and devices. This data will be possibly managed, processed, and analyzed, as part of the managing procedure [44]. There are various methods of storing and processing the data (e.g., centralized/distributed or locally/remotely). The common state-of-the-art method is by using ***cloud computing*** which is a new way of utilizing shared computer processing resources. Cloud computing and storage solutions provide users and enterprises with various capabilities to store and process their data in data centers that may be located far from the user (across the world). It integrates with IoT in order to provide the computing power required for big data processing [34]. Moreover, cloud computing provides the customers with infrastructure, platform, software, and sensor network, as services [45–51]. These services are guaranteed by the providers to have the reliability and performance stated in the agreements, which can be essential for the energy management platform.

Cloud computing might be suitable for applications without delay-sensitive operations. However, in some applications, response time and latency are critical factors for delay-sensitive devices. The timing requirements have to be met in order to have a proper and efficient operation (e.g., energy management) [52–54]. Therefore, cloud computing in IoT may be lacking the required performance, especially when increasing the number of existing devices may exacerbate the problem.

Fog computing may be the current best solution to this problem. As shown in Fig. 7.2, the paradigm of cloud computing is moved further to the edge of network (intelligence at the edge) and closer to the devices. It adds an intermediary layer to the platform providing the IoT with the capability of pre-processing the data while meeting the low-latency requirements [39]. It uses multiple on-the-edge (near-user) devices to carry out a substantial amount of storage, communication, control, and processing capability. The fog computing offers other benefits which might be useful in some areas and domains such as in energy management [39]:

- Deploying fog computing can help developers quickly implement the platform and applications by offering service-oriented architectures (SOAs) in the way each customer needs.
- Developers can use their own security policy and analyze sensitive data locally eliminating transmitting to the cloud for better privacy.

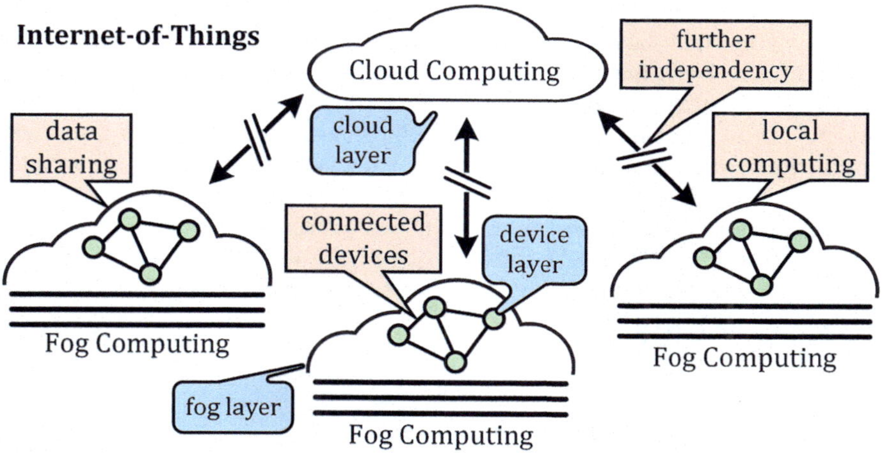

Fig. 7.2 Implementing fog computing as an intermediary layer to cloud computing in Internet-of-Things (IoT) platform

- Conserve network bandwidth by processing selected partial data locally and help the cloud by decreasing the transmission rate which may reduce the operating and infrastructure cost of the cloud.

7.4 Control-as-a-Service

The energy management platform can be used for any type of buildings and various domains of operation, e.g., home or microgrid. The energy management may have various purposes like: (1) monitoring and metering the power consumption of each device, e.g., home power consumption; (2) managing the energy consumption by controlling the devices efficiently, e.g., intelligent lighting, Electric Vehicle (EV) charger [55, 56], Heating, Ventilation, and Air Conditioning (HVAC) management, etc. The energy management platform is a system of systems. As in energy management system, the platform should be adaptive and scalable regardless of the system size. It also should provide the devices with the performance, interoperability, and interactivity features in order to help them transfer and process the data generated in an adequate response time. The energy management—as a control application—can utilize the fog computing platform to provide the needed scalable connectivity and interoperability as well as the low-latency computation power which is at the edge of the network.

However, a more cost-effective and customizable design is required for increasing the chance of penetrating into the market. This can be achieved by designing the controlling functionalities as services on the platform. SOA on top of fog computing is the answer for implementing the Control-as-a-Service (CaaS). The SOA infrastructure provides a scalable hardware, software, and communication architecture which can meet all-the-above-mentioned requirements for an energy management platform. The basic fundamental principles of SOA are independent of vendors, products, infrastructure, and technologies. A service is a discrete unit of functionality that can be accessed remotely and acted upon and updated independently, such as energy management in different levels of the power grid. SOA aims to allow users and developers to combine selective functionalities to form applications which are built purely from existing services and combining them in an ad hoc manner. The hardware (see Sect. 7.4.1), software (see Sect. 7.4.2), and communication (see Sect. 7.4.3) architectures of this platform are defined and integrated (see Sect. 7.4.4) in detail for a CPES.

7.4.1 Hardware Architecture

The hardware of the energy management platform comprises of multiple devices, depending on its operating domain, e.g., home or microgrid. These devices may be categorized as following, based on their functionalities:

1. **Connecting:** these devices provide the connectivity capability among the existing and compatible devices. The wires, sockets, and antennas are considered as the connecting devices. The speed of these connections can define the performance of the platform in terms of latency and bandwidth.
2. **Gateway:** various devices may communicate differently using different standards and protocols, e.g., ZigBee, Bluetooth, and Ethernet [57]. The gateway devices establish a compatible connectivity and interface between these devices if required.
3. **Sensor:** the energy management system needs to monitor the environment for timely changes, e.g., weather, light, energy price, etc. The sensors may digitize the analog signal generated by the environment. The sampling rate and precision of the sensors are defined by the application requirements.
4. **Actuator:** the energy management system may decide to configure multiple devices according to environment changes, in order to optimize a variable, e.g., energy consumption. These configurable devices are considered as actuators which may be locally or remotely controlled. Actuators abstract any controllable device which can whether be as sophisticated as a complex embedded system (HVAC) or as simple as a switch (light).
5. **Computing:** the devices that store, process, and analyze the data in the system. They may also implement sophisticated control algorithms to configure the actuating devices. Their processing and storing capabilities are defined by the application requirements which can be as small as the processor of a router.

All the devices existing in an energy management platform are considered as connected nodes in the network of the IoT. Moreover, multiple devices or nodes might be grouped as a subsystem in order to perform a single function. For instance, in an

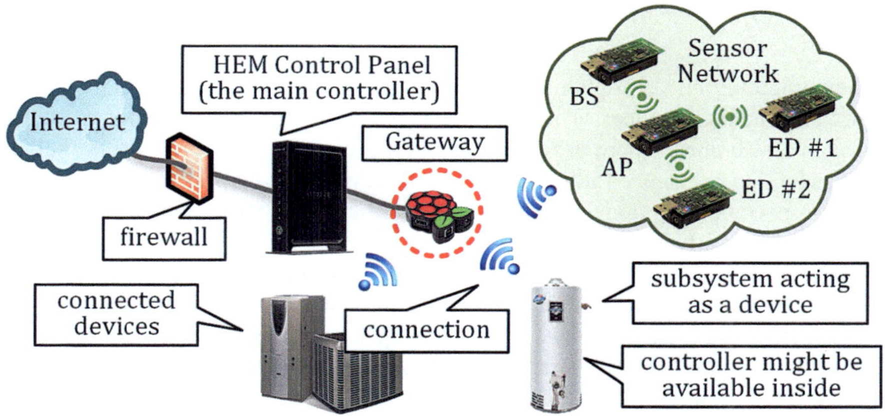

Fig. 7.3 Illustrating the hardware architecture of the fog computing platform for a home energy management (HEM) [10]

HEM platform, multiple smart devices, e.g., HVAC, are available to be monitored (sensed) and controlled (actuated) in order to reduce the total energy consumption. The essential computing device which handles all the monitoring and controlling tasks is called the **HEM control panel** (a router). Its main job is to discover and monitor different devices in the platform dynamically, handle different time-flexible load requests, and trigger and command the devices accordingly based on the algorithms implemented, in order to optimize a variable. To manage the energy consumption of a home, the HEM platform might need to retrieve information about the environment of the home. The physical world condition may be monitored by multiple sensors (e.g., temperature, humidity, light, etc.) in different places of the home. Moreover, different types of sensors may be implemented on a single device (e.g., TelosB mote module). Also, a smart device may have a computing node, implemented inside of it. This computing device, **device control panel**, enables the user or HEM control panel with the capability of configuring the device or a subsystem (see Fig. 7.3), in lower level and much more detail than the HEM control panel may be capable of. This fog-enabled IoT platform can be expanded further for a wider, more complex, and heterogeneous system.

Figure 7.3 shows the hardware architecture used for communication and computation in the fog computing platform. As you see in the figure, there may exist multiple devices with the processing capability which optimize their local decisions at device level. These enable the intelligence at the edge for the fog computing platform. They may retrieve global (higher level) information from the main control panel (HEM Control Panel) for further decision-making and adaptation. Moreover, the gateway is required for compatible interface may be eliminated, if the device or subsystem is capable of communicating with the platform directly.

7.4.2 Software Architecture

The computing nodes implement controllers to gather, store, process, and analyze data and manage devices. These computing devices can be as simple as routers available at homes. There are open-source and user-configurable routers that run a distribution of Linux [58] on a MIPS processor. Developers can utilize these routers as computing nodes to easily program the controllers, compile, and run the algorithms directly on the routers.

Moreover, the sensor devices, e.g., TelosB mote modules are compatible with TinyOS [59] which is an open-source operating system (available in [60]) designed for low-power wireless sensors. Therefore, the algorithms implemented for monitoring, commanding the sensors, and receiving data from them may be programmed easily and dynamically on the sensors. This flexibility may help the developers to program the sensors based on their requirements. In other words, different routing and discovery algorithms for different kinds of sensors (e.g., temperature, humidity, and light) may be implemented in different scenarios. Hence, the sensors can also be treated as the computation nodes for local decision-making alongside the communicating purposes.

The control panels for each subsystem manage their devices in their own network through a predefined protocol. However, all the subsystems and devices connected in the main network need to follow a unique protocol defined by the HEM control panel (e.g., ZigBee). Therefore, the communications between devices are built around Devices Profile for Web Services (DPWS) from the Web Services for Devices (WS4D). Also, it relies on SOAP-over-UDP, SOAP, WSDL, and XML Schema. This protocol stack may be used for sending secure messages from device to device in a heterogeneous platform. The developers may utilize the WS4D-gSOAP toolkit available in [61] to implement the services needed for each device. Devices may host various DPWS-compliant services using:

1. **WS-Addressing:** provides an addressing mechanism for web services as well as messages in a transport-neutral matter.
2. **WS-Discovery:** provides a way to make the services discoverable by leveraging a protocol based on IP multicast.
3. **WS-MetadataExchange:** defines different data types and the operations to receive a metadata.
4. **WS-Transfer:** is used to transfer the metadata and is very similar to HTTP.
5. **WS-Eventing:** describes a protocol which allows web services to subscribe to or accept subscriptions for event notification messages.

Since the web services are platform-agnostic, using SOA, the platform maintains its flexibility of adding new devices, discovering devices and their respected services, synchronization between devices, and transferring structured data between them. Also, in an HEM platform, the devices or the controllers required by a user may be added and managed as a service (CaaS). The devices broadcast the functions they already have implemented to other devices, especially the main

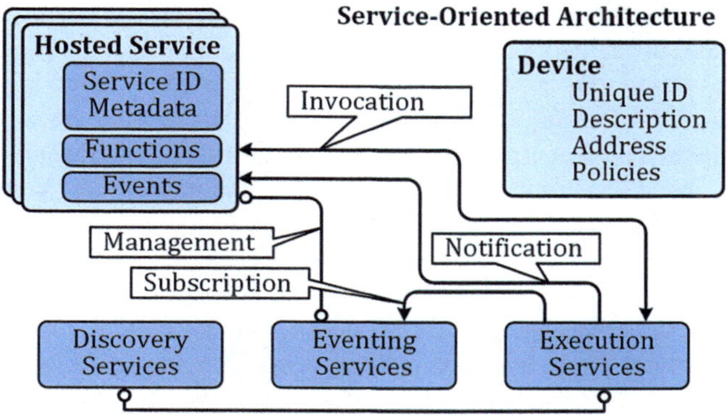

Fig. 7.4 Demonstration of the software architecture implemented as a service-oriented architecture (SOA) [10]

control panel. Hence, they may call these functions within the fog. Moreover, to establish synchronization capability and interoperability between devices in the network, e.g., HEM control panel, the device may host "eventing" web services. In this way, the devices will be notified of any changes or events related to other devices if they have subscribed to those events. Figure 7.4 shows how these services are implemented and configured in the platform in order to provide communication and certain functionalities required for the devices.

7.4.3 Communication Architecture

The devices in the platform communicate and can use the existing network adapters and interfaces. Deploying the current communication infrastructure is one of the main factors of the fog computing in order to make an affordable platform. For instance, the home routers in the HEM platform may have Wireless, Ethernet, Bluetooth, Universal Serial Bus (USB), etc., according to their specifications. Hence, they can be leveraged for communication between devices. Since a unique protocol (e.g., ZigBee) should be used for the main network, these interfaces need to be converted using gateway devices for the desired protocol, if necessary.

According to the platform requirement, a large sensor network can be implemented. For instance, multiple sensor nodes may be connected together to form a wireless sensor network for better monitoring the environment. The sensor devices, e.g., TelosB mote modules, may use the standards-based, low-power wireless technology—ZigBee—to communicate with each other. Moreover, the standard used for the communication between different nodes may be Smart Energy Profile (SEP 2.0), which is a standard for IP-based control for HEM and it is supported by these devices, but this profile is not ready yet (standard, protocol stack, and hardware). Furthermore, the flexibility of the ZigBee networks can handle about 65,535 devices in a network. All the devices are tagged using a unique Identification (ID) number. These IDs are hard coded into the sensor devices while programming and maybe used for routing and transferring data in the same network. Also, the services help the HEM control panel to discover new devices added via plug-n-play. The HEM control panel will verify and authorize the new devices connected to the HEM platform.

To further extend the range that sensor network supports, a two-level hierarchical network can be utilized. In other words, some of the sensor devices are programmed to only sense, which are called End Devices (ED) and may be placed in different corners of a room. Then, a sensor node may be programmed to act as an Access Point (AP) and connects to all the end devices in that room. Also, in a higher level, another sensor device may act as a Base Station (BS) which connects to multiple access points in all the rooms. The end devices transfer the data through their access points and then to the base station. The number of access points connected to each base station and the number of end devices connected to the access points are the variables which are adjusted based on the structure of the building and the requirements. The base station sends all the data gathered from all end devices to the HEM control panel directly or indirectly (using gateway), at the final stage. The gateway used here is a Raspberry Pi. The sensor nodes may communicate with a Raspberry Pi using serial connection over ZigBee standard, and then the Raspberry Pi connects to the HEM control panel through Ethernet. However, if the router has the capability of communicating via ZigBee directly, or the router had the compatible driver to control the sensor module through USB, the Raspberry Pi would be eliminated. Figure 7.5 demonstrates the protocols and web services used for the communications between devices.

Furthermore, the user may interface with the platform using the web pages designed for the control panel of each device. The HEM control panel interface is

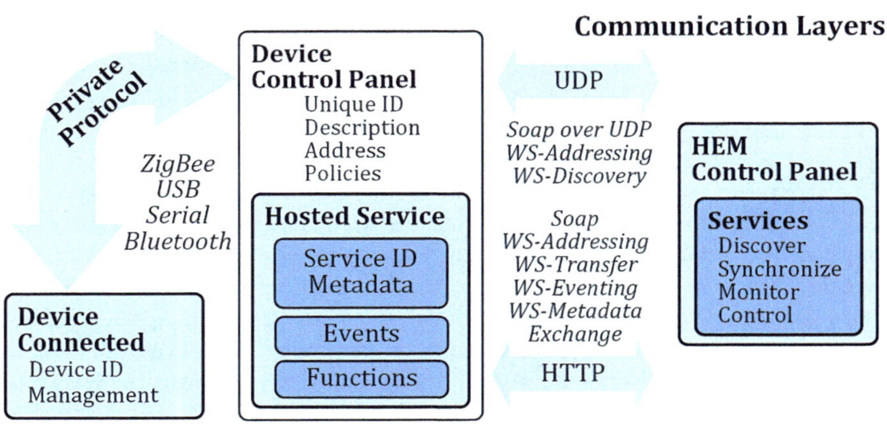

Fig. 7.5 Layers of the communication architecture used in the fog computing platform [10]

setup on the main router. However, the interfaces for other devices may be provided by the vendors of the devices. Moreover, Asynchronous JavaScript and XML (AJAX) technique may be utilized to retrieve the information from the devices and view it on the web page and trigger different functions and services implemented on the control panel in order to process data. Furthermore, connecting the HEM control panel to the Internet enables the user to monitor and control the home locally or remotely through internet.

7.4.4 Integration of Architectures

As we have explained in the last three subsections, the flexibility and low-cost infrastructure helps developers to add any device, sensor, actuator, and their respected services. The functions and controllers are implemented as services in the fog computing platform using SOA. The services of the devices are defined using WS4D to be able to communicate with the routers and each other even among heterogeneous devices. On the other hand, if a subsystem designed for an application-specific purpose, needs to be added to the platform, regardless of its own protocol and architecture, it needs to be compatible with the protocol stack used in the HEM platform, or it can become compatible by using gateway devices. Hence, the computation can be done on the devices close to the network even the sensor nodes for the local decision-making. This will eliminate the need of transferring data to the cloud for computation. Furthermore, the scalability and adaptability of the platform for other areas and applications become easier in this platform.

7.5 Residential Cyber-Physical Energy System

The implementation of CaaS on top of fog computing platform is applied to two prototypes of HEM and microgrid-level energy management to demonstrate its advantages for different domains.

7.5.1 Home Energy Management

The HEM platform is implemented for one home as shown in Fig. 7.6. In the home, multiple smart devices such as: HVAC, water heater, and EV charger are implemented. Each device is monitored and controlled by the HEM control panel. Also, the devices have their own control panels to monitor their status and set their configurations. In this prototype, all the devices are software modeled; however, in the real-world implementation, the control panels of the devices will be provided by their vendors.

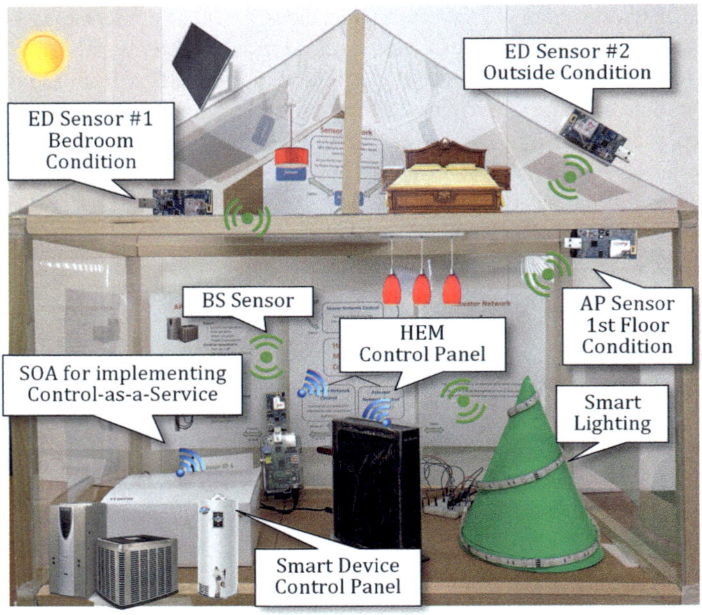

Fig. 7.6 Prototype demonstration of an HEM over the fog computing platform [10]

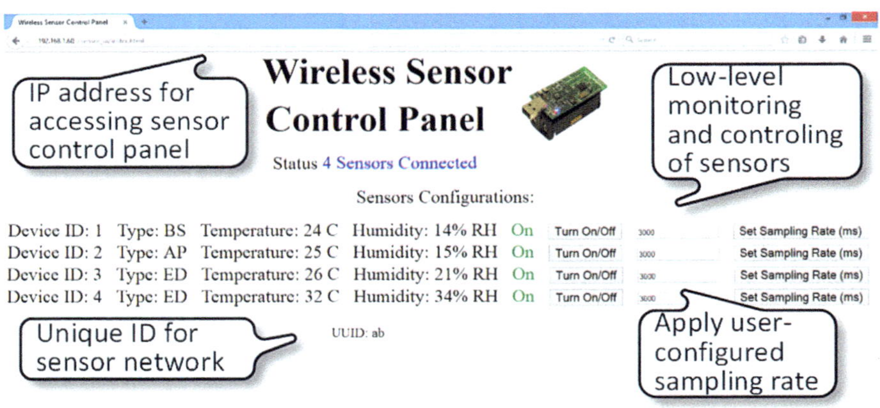

Fig. 7.7 Sensor network control panel for managing sensors' configurations [10]

The home is being monitored by a network of four sensor devices (TelosB mote modules). The sensor network is defined as a subsystem inside the platform (system of systems). They sample the temperature, humidity, and lighting inside and outside of the home. Also, different types of commands to enable/disable the sensors or set the configurations, e.g., sampling rate for the sensors, may be sent by the sensor control panel to each device by specifying its ID number (see Fig. 7.7).

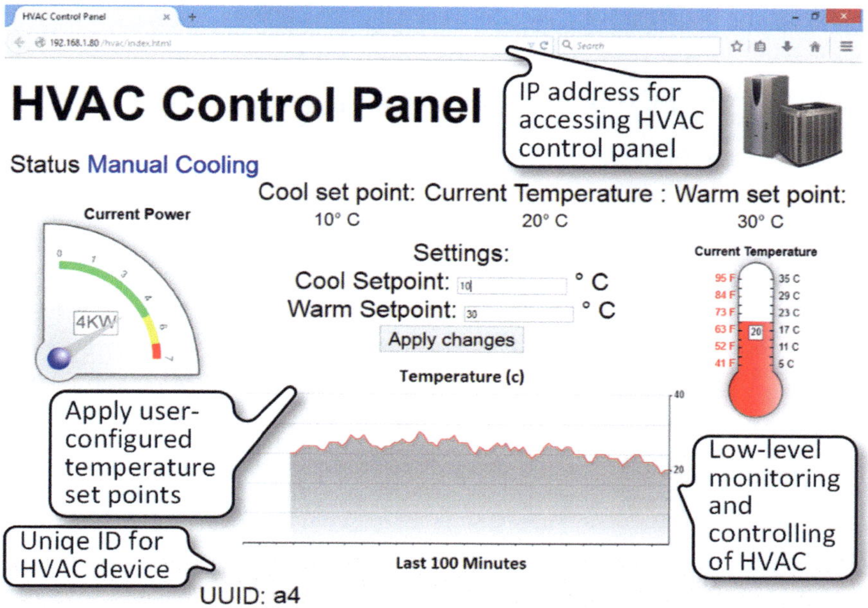

Fig. 7.8 Heating, ventilation, and air conditioning (HVAC) control panel for monitoring and controlling HVAC power [10]

In the HEM platform, an HVAC controller has been implemented as a service. The controller manages the HVAC in order to reduce its power consumption. It gathers information such as the temperature set points configured by the user, the Demand Response (DR) signal sent by the transformer, the current room temperature, and HVAC operation mode (e.g., cooler or heater). It adjusts the temperature set points and the status of the HVAC (e.g., on or off) according to the decision-making algorithm. Initially, the threshold for turning on/off the HVAC is defined. The controller turns off/on the HVAC system based on the operation mode and the temperature difference between the room and the user preference. The adjusted HVAC status and the temperature set points are sent to the HVAC (in the device level), as shown in Fig. 7.8.

EV charger controller has been also implemented as a service in the HEM platform. The controller schedules the charging time of the battery in order to reduce the electricity cost while meeting the departure time of the user. It receives user-specified departure time, the current time of the day, and the current battery status. Then, it adjusts the EV charging rate. The controller has the information about the electricity pricing policy. Hence, it schedules the charging process over the off-peak hours first. Then, if more energy is required, more charging will be allocated during the on-peak hours. The charge rate is assigned based on the current time.

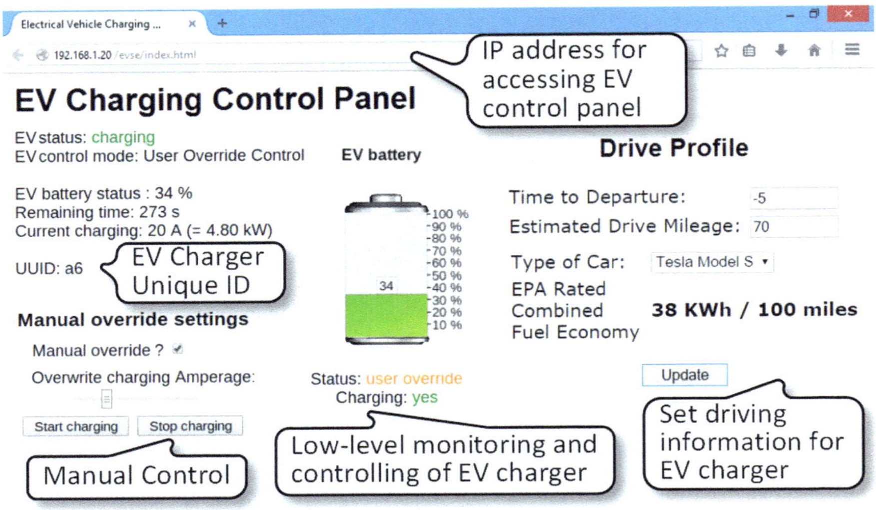

Fig. 7.9 EV control panel for monitoring and controlling battery charging [10]

In the HEM, services like managing sensor network, efficient lighting, smart EV charging, and smart HVAC controlling are added to the energy management system. The device-level controllers for each of the services are implemented separately in the HEM fog computing platform. The HEM control panel views the current devices connected to the platform. Through HEM platform, the user may turn on/off each device (see Fig. 7.10). Figure 7.7 shows the user interface for managing the sensor network in the home. The user may check the temperature and humidity in different parts of the home. Using smart EV charging, the EV charger will efficiently decide when and how to charge the EV such that it does not violate power consumption restrictions while meeting the departure time specified by the user (see Fig. 7.9). The HVAC consumption is mainly dependent on the temperature set points adjusted by the user. By monitoring the temperature in different places of the home and knowing the energy price, the HVAC controller may efficiently decide on the temperature set points (see Fig. 7.8). This process may reduce the power consumption of the HVAC while maintaining the home temperature within the defined range.

Although in this example, the main control panel is responsible for monitoring variables and adjusting different temperature set points, it can get more complex. The HEM control panel can implement a sophisticated Supervisory Control and Data Acquisition (SCADA) controller. It enables the HEM to monitor and jointly optimize the controlling variables of all the devices (Fig. 7.10).

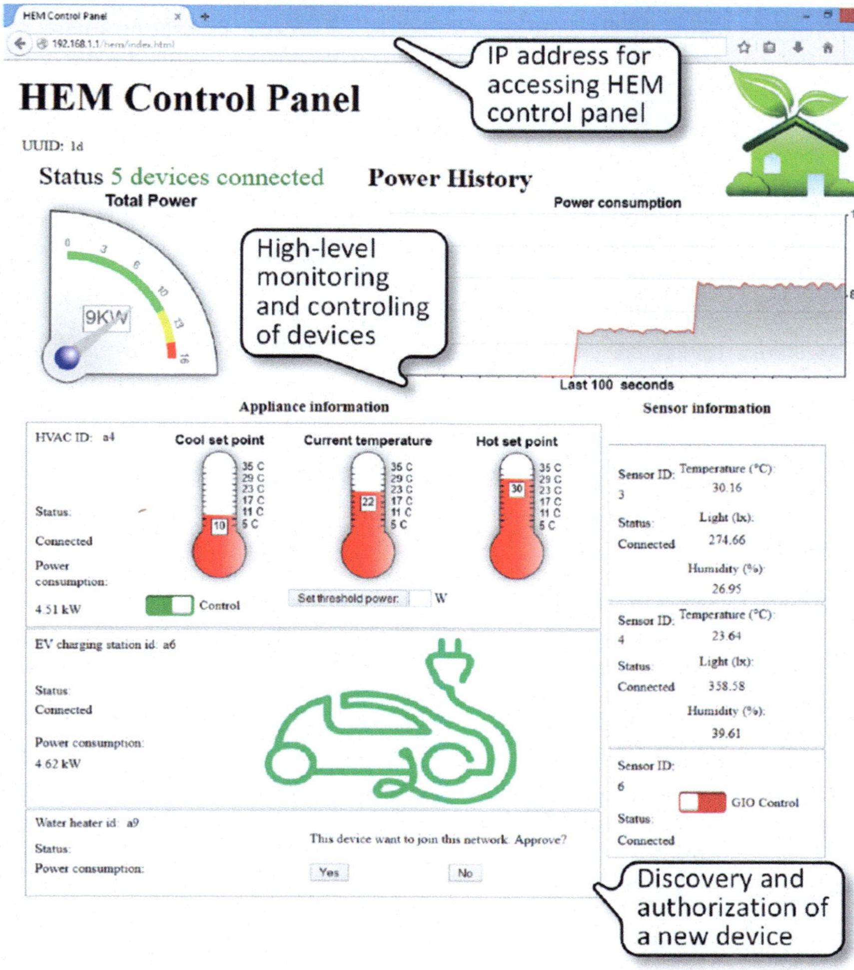

Fig. 7.10 HEM control panel for monitoring and controlling smart devices [10]

7.5.2 *Microgrid-Level Energy Management*

The microgrid-level energy management platform comprises three homes connected to a transformer. A control panel is implemented in the transformer to monitor and manage the power consumption of each home. The transformer-level control panel monitors the load of each home and may decide to send commands to their HEM, in order to reduce their power consumption by a specific value (DR). The power reduction may prevent overloading and overheating the transformer and may

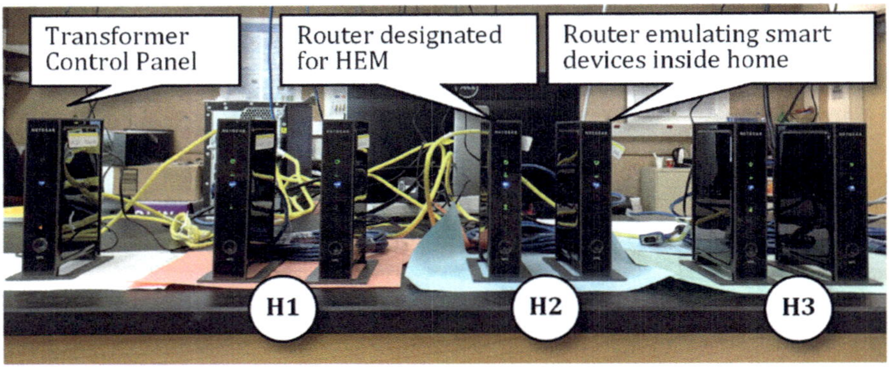

Fig. 7.11 Microgrid-level energy management prototype [10]

improve its efficiency and longevity. The transformer control panel is implemented in one router. Other three homes and their emulated smart devices are implemented in other six routers (see Fig. 7.11).

In the microgrid-level energy management platform, a transformer management has been implemented as a service. The controller monitors the homes connected to the transformer in order to prevent overloading the transformer. It receives information about the array of the homes connected and the current load on the transformer. The controller checks whether the transformer is overloading. In case of overload, a DR signal is sent to each home which consumes more than the threshold. The controllers for the service are implemented in the microgrid-level fog computing platform. Figure 7.12 shows the user interface for monitoring the power consumption of each home. The user may define certain power threshold for each home. The transformer current condition and the total load on it may be monitored. Using this information, the transformer will send DR signals to the home which have violated the threshold so that they may reduce their consumption. Also, the control panel may decide where to get the power from (see Fig. 7.12).

To demonstrate and experiment the features of the energy management over fog computing platform, the complete working prototypes of the case study demonstrations are presented in [24]. Both energy management systems have provided the required features for an affordable platform by implementing the cost-effective SOA over fog computing using off-the-shelf instruments. Table 7.1 summarizes the technologies used to provide these features as in the example of the residential CPES.

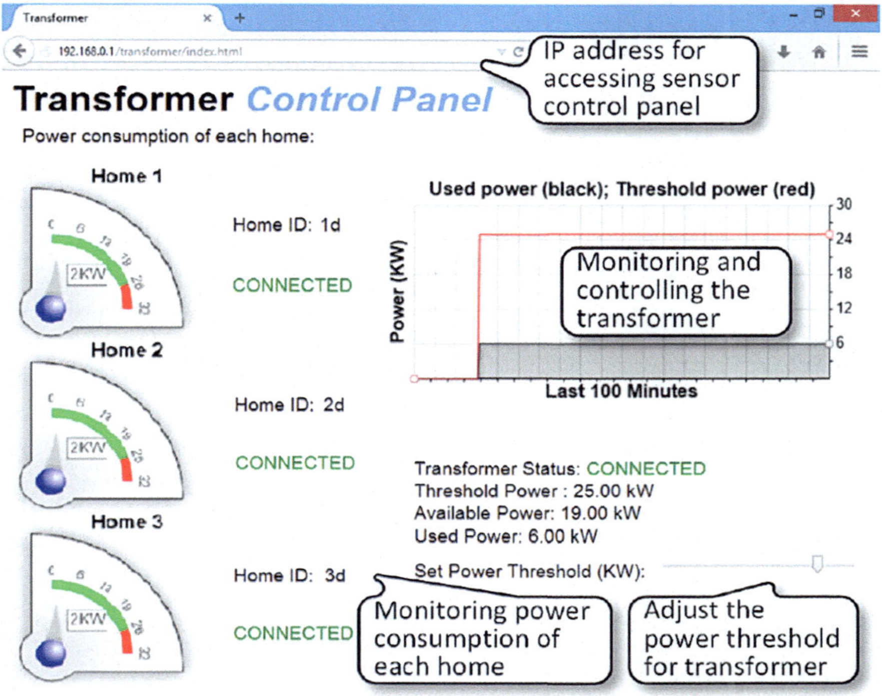

Fig. 7.12 Transformer-level control panel for monitoring and controlling power of each home [10]

Table 7.1 Summary of the technologies used to implement energy management platform [10]

Feature	Used technology
Interoperability	WS-Addressing and WS-Transfer enable the devices to communicate with each other
Interactivity	WS-Eventing and WS-Metadata Exchange enable devices to synchronize each other periodically
Flexibility	Open hardware/software used for implementing any application or adding new devices
Scalability	Multiple devices can be connected in a shared network with their own unique IDs
Ease of deployment	Control panel web pages leverage HTML, AJAX, and Java scripting in order to provide user-friendly interfaces
Open architecture	Raspberry Pi is used as gateway and Linux-based routers and Tiny OS-based TelosB mote sensors are used for connecting and computing
Plug-n-play	SOAP-over-UDP is used for discovery and authentication of new devices added to network
Local/remote access	IP addresses are designated for each device and they are also connected to the Internet
Heterogeneity abstraction	Service-oriented architecture is used to abstract the hardware and communication differences

References

1. Berkeley, Microgrids at Berkeley lab (2015) [Online]. Available: der.lbl.gov
2. M.A. Al Faruque, RAMP: impact of rule based aggregator business model for residential microgrid of prosumers including distributed energy resources, in *IEEE PES Innovative Smart Grid Technologies Conference (ISGT)*, pp. 1–6, 2014
3. K. Vatanparvar, M.A. Al Faruque, Design space exploration for the profitability of a rule-based aggregator business model within a residential microgrid. IEEE Trans. Smart Grid (TSG) **6**(3), 1167–1175 (2015)
4. S. Shao, M. Pipattanasomporn, S. Rahman, Challenges of PHEV penetration to the residential distribution network. IEEE Power and Energy Society General Meeting, 2009
5. S. Shao, T. Zhang, M. Pipattanasomporn, S. Rahman, Impact of TOU rates on distribution load shapes in a smart grid with PHEV penetration, in *IEEE PES Transmission and Distribution Conference and Exposition: Smart Solutions for a Changing World*, 2010
6. M. Jafari, Optimal energy management in community micro-grids, in *IEEE PES Innovative Smart Grid Technologies (ISGT)*, pp. 1–6, 2012
7. Department of Energy (DOE), Buildings energy data book (2014) buildingsdatabook.eren.doe.gov/TableView.aspx?table=1.1.3
8. U.S. Department of Energy Information Administration, Washington, DC. Electric power monthly. http://www.eia.gov/electricity/monthly/current_year/april2014.pdf. April, 2014. [June, 2014]
9. B. Bolin, B. Doos, R. Warrick, J. Jaeger, *The Greenhouse Effect, Climatic Change, and Ecosystems*, 29 (Wiley and Sons (SCOPE), New York, 1986)
10. M.A. Al Faruque, K. Vatanparvar, Energy management-as-a-service over fog computing platform. IEEE Internet Things J. **3**(2), 161–169 (2016)
11. United States Environmental Protection Agency, Integrated energy policy report update (2004). Internet: http://www.energy.ca.gov/reports/CEC-100-2004-006/CEC-100-2004-006CMF.PDF [June, 2014]
12. United States Environmental Protection Agency. Clean power plan. Internet: http://www2.epa.gov/sites/production/files/2014-06/documents/20140602ria-clean-power-plan.pdf. June, 2014, [June, 2014]
13. California Energy Emission, Building energy efficiency program. www.energy.ca.gov/title24, 2014
14. Department of Energy, US Department of energy strategic plan. energy.gov/sites/prod/files/2011_DOE_Strategic_Plan_.pdf, 2014
15. S. Teleke, M.E. Baran, S. Bhattacharya, A.Q. Huang, Rule-based control of battery energy storage for dispatching intermittent renewable sources. IEEE Trans. Sustainable Energy **1**(3), 117–124 (2010)
16. P. Siano, Demand response and smart grids' A survey. Renew. Sust. Energ. Rev. **30**, 461–478 (2014)
17. J. Aghaei, M.-I. Alizadeh, Demand response in smart electricity grids equipped with renewable energy sources: a review. Renew. Sust. Energ. Rev. **18**, 64–72 (2013)
18. M.A. Al Faruque, L. Dalloro, S. Zhou, H. Ludwig, G. Lo, Managing residential-level EV charging using network-as-automation platform (NAP) technology, pp. 1–6, 2012
19. M. Pipattanasomporn, M. Kuzlu, S. Rahman, An algorithm for intelligent home energy management and demand response analysis. IEEE Trans. Smart Grid **3**(4), 2166–2173 (2012)
20. M.A. Al Faruque, F. Ahourai, A model-based design of cyber-physical energy systems, in *Asia and South Pacific Design Automation Conference (ASP-DAC)*, pp. 97–104, 2014
21. DER Group at LBNL, "WebOpt". der.lbl.gov/News/3rd-release- distributed-energy-resources-der-web-optimization-tool-webopt, 2014
22. T.H. Morris, A.K. Srivastava et al., Engineering future cyber-physical energy systems: challenges, research needs, and roadmap, in *North American Power Symposium (NAPS)*, pp. 1–6, 2009

23. S. Karnouskos, Cyber-physical systems in the SmartGrid, in *IEEE International Conference on Industrial Informatics (INDIN)*, pp. 20–23, 2011
24. K. Vatanparvar, M.A. Al Faruque, Demo abstract: energy management as a service over fog computing platform, in *ACM/IEEE International Conference on Cyber-Physical Systems (ICCPS)*, pp. 248–249, 2015
25. K. Vatanparvar, Q. Chau, M.A. Al Faruque, Home energy management as a service over networking platforms, in *IEEE PES Conference on Innovative Smart Grid Technologies (ISGT)*, 2015
26. U.S. Department of Energy Efficiency and Renewable Energy Golden Service Center. Financial Assistant Funding Opportunity Announcement. 28 March 2013. DE-FOA-0000822: "Turn key" open source software solutions for energy management of small to medium sized buildings (2014)
27. F. Jammes, H. Smit, Service-oriented paradigms in industrial automation. IEEE Trans. Ind. Inf. (2005). doi:10.1109/TII.2005.844419
28. P. Palensky, D. Dietrich, Demand side management: demand response, intelligent energy systems, and smart loads. IEEE Trans. Ind. Inf. **7**, 381–388 (2011)
29. D. Bian, M. Kuzlu, M. Pipattanasomporn, S. Rahman, Assessment of communication technologies for a home energy management system, in *IEEE PES Innovative Smart Grid Technologies Conference (ISGT)*, pp. 1–5, 2014
30. M. Rahman, M. Kuzlu, M. Pipattanasomporn, S. Rahman, Architecture of web services interface for a home energy management system, in *IEEE PES Innovative Smart Grid Technologies Conference (ISGT)*, pp. 1–5, 2014
31. D.-M. Han, J.-H. Lim, Design and implementation of smart home energy management systems based on ZigBee. IEEE Trans. Consum. Electron. **56**, 1417–1425 (2010)
32. Y.-S. Son, T. Pulkkinen, K.-D. Moon, C. Kim, Home energy management system based on power line communication. IEEE Trans. Consum. Electron. **56**, 1380–1386 (2010)
33. S. Katipamula, R.M. Underhill, J.K. Goddard, D. Taasevigen, M. Piette, J. Granderson, R.E. Brown, S.M. Lanzisera, T. Kuruganti, Small-and medium-sized commercial building monitoring and controls needs: a scoping study. Technical Report, Pacific Northwest National Laboratory (PNNL), Richland, WA (2012)
34. X. Ye, J. Huang, A framework for cloud-based smart home, in *Proceedings of International Conference on Computer Science and Network Technology (ICCSNT)*, vol. 2, pp. 894–897, 2011
35. J. LaMarche, K. Cheney, S. Christian, K. Roth, Home energy management products & trends. Fraunhofer Center for Sustainable Energy Systems (2011)
36. C. Angulo, R. Téllez, Distributed intelligence for smart home appliances. Tendencias de la minería de datos en España. Red Española de Minería de Datos (2004)
37. M. Skubic, G. Alexander, M. Popescu, M. Rantz, J. Keller, A smart home application to eldercare: current status and lessons learned. Technol. Health Care **17**(3), 183–201 (2009)
38. S.Y. Chen, C.F. Lai, Y.M. Huang, Y.L. Jeng, Intelligent home-appliance recognition over IoT cloud network, in *9th International Wireless Communications and Mobile Computing Conference (IWCMC)*, pp. 639–643, 2013
39. F. Bonomi, R. Milito, J. Zhu, S. Addepalli, Fog computing and its role in the internet of things, in *Proceedings of the First Edition of the MCC Workshop on Mobile Cloud Computing*, pp. 13–16, 2012
40. A.-M. Rahmani, N.K. Thanigaivelan, T.N. Gia, J. Granados, B. Negash, P. Liljeberg, H. Tenhunen, Smart e-health gateway: bringing intelligence to internet-of-things based ubiquitous healthcare systems, in *IEEE Consumer Communications and Networking Conference (CCNC)*, pp. 826–834, 2015
41. Z. Sheng, S. Yang, Y. Yu, A. Vasilakos, J. McCann, K. Leung, A survey on the IETF protocol suite for the internet of things: standards, challenges, and opportunities. IEEE Wirel. Commun. **20**, 91–98 (2013)
42. Z. Yan, P. Zhang, A.V. Vasilakos, A survey on trust management for Internet of Things. J. Netw. Comput. Appl. **42**, 120–134 (2014)

43. Q. Jing, A.V. Vasilakos, J. Wan, J. Lu, D. Qiu, Security of the Internet of Things: perspectives and challenges. Wirel. Netw. **20**, 2481–2501 (2014)
44. A. Zaslavsky, C. Perera, D. Georgakopoulos, Sensing as a service and big data, arXiv preprint arXiv:1301.0159 (2013)
45. L. Wang, F. Zhang, J. A. Aroca, A. V. Vasilakos, K. Zheng, C. Hou, D. Li, Z. Liu, GreenDCN: a general framework for achieving energy efficiency in data center networks. IEEE J. Sel. Areas Commun. **32**, 4–15 (2014)
46. S. Patidar, D. Rane, P. Jain, A survey paper on cloud computing, in *Proceedings of 2nd International Conference on Advanced Computing and Communication Technologies (ACCT)*, pp. 394–398, 2011
47. Q. Duan, Y. Yan, A.V. Vasilakos, A survey on service-oriented network virtualization toward convergence of networking and cloud computing. IEEE Trans. Netw. Serv. Manage. **9**, 373–392 (2012)
48. M.R. Rahimi, N. Venkatasubramanian, A.V. Vasilakos, MuSIC: mobility-aware optimal service allocation in mobile cloud computing, in *IEEE International Conference on Cloud Computing, CLOUD*, pp. 75–82, 2013
49. G. Fortino, G. Di Fatta, M. Pathan, A.V. Vasilakos, Cloud-assisted body area networks: state-of-the-art and future challenges. Wirel. Netw. **20**, 1925–1938 (2014)
50. L. Wei, H. Zhu, Z. Cao, X. Dong, W. Jia, Y. Chen, A.V. Vasilakos, Security and privacy for storage and computation in cloud computing. Inf. Sci. **258**, 371–386 (2014)
51. Q.Z. Sheng, X. Qiao, A.V. Vasilakos, C. Szabo, S. Bourne, X. Xu, Web services composition: a decade's overview. Inf. Sci. **280**, 218–238 (2014)
52. A. Copie, T.-F. Fortis, V.I. Munteanu, Benchmarking cloud databases for the requirements of the internet of things, in *Information Technology Interfaces (ITI)*, pp. 77–82, 2013
53. F. Xu, F. Liu, H. Jin, A.V. Vasilakos, Managing performance overhead of virtual machines in cloud computing: a survey, state of the art, and future directions. Proc. IEEE **102**, 11–31 (2014)
54. L. Wang, F. Zhang, A.V. Vasilakos, C. Hou, Z. Liu, Joint virtual machine assignment and traffic engineering for green data center networks. ACM SIGMETRICS Perform. Eval. Rev. **41**, 107–112 (2014)
55. M.A. Al Faruque, L. Dalloro, S. Zhou, H. Ludwig, G. Lo, Managing residential-level EV charging using network-as-automation platform (NAP) technology, in *IEEE International Electric Vehicle Conference (IEVC)*, 2012
56. M.A. Al Faruque, A. Canedo, Intelligent and collaborative embedded computing in automation engineering, in *Proceedings of the Conference on Design, Automation and Test in Europe*, pp. 344–345, 2012
57. P. Baronti, P. Pillai, V.W. Chook, S. Chessa, A. Gotta, Y.F. Hu, Wireless sensor networks: a survey on the state of the art and the 802.15.4 and ZigBee standards. Comput. Commun. **30**, 1655–1695 (2007)
58. DD-WRT. Open source firmware for routers (2014). dd-wrt.com/site/index
59. P. Levis, S. Madden, J. Polastre, R. Szewczyk, K. Whitehouse, A. Woo, D. Gay, J. Hill, M. Welsh, E. Brewer, D. Culler, TinyOS: an operating system for sensor networks, in *Ambient Intelligence* (Springer, Berlin, 2005), pp. 115–148
60. TinyOS Open source operating system for low-power wireless devices. github.com/tinyos. 2014
61. Web services for devices "(WS4D)". ws4d.e-technik.uni-rostock.de. 2014

Chapter 8
Leveraging Fog Computing for Healthcare IoT

Behailu Negash, Tuan Nguyen Gia, Arman Anzanpour, Iman Azimi, Mingzhe Jiang, Tomi Westerlund, Amir M. Rahmani, Pasi Liljeberg, and Hannu Tenhunen

8.1 Introduction

The Internet of Things (IoT) is already around us. We can see it in many areas of our daily life; from wearable fitness tracking gadgets, smart home appliances, to smart self-driving cars [1, 2]. Healthcare is expected to be among the domains that will be remodeled through IoT by enhancing its penetration and lowering the service cost [3, 4]. IoT offers potential to uninterrupted and reliable remote monitoring due to its ubiquitous nature while allowing freedom of movement for individuals. Activity tracking and following up of the heart rate and calorie intake are some of the application areas of commercially available IoT devices. In professional medical environments also, the quality of healthcare services can be enhanced by automating patient monitoring [5–7]. This enables a coherent healthcare system at home and in the hospital through the cloud [4, 8]. It is predicted that the current hospital-centered practice of medical care will be balanced by its home-based counterpart in 2020. This progress is expected to reach home-centered approach in the next decade[9]. To support such a shift and scaling, new computing approaches need to be developed.

Integration of currently available fragmented solutions towards all inclusive healthcare is a critical requirement, and at the same time potential, in IoT. This integration allows the synchronization of data from wearable and implantable

B. Negash (✉) • T.N. Gia • A. Anzanpour • I. Azimi • M. Jiang • T. Westerlund • P. Liljeberg • H. Tenhunen
University of Turku, Turku, Finland
e-mail: behneg@utu.fi; tunggi@utu.fi; armanz@utu.fi; imaazi@utu.fi; mizhji@utu.fi; tovewe@utu.fi; pakrli@utu.fi; hannu@kth.se

A.M. Rahmani
Department of Computer Science, University of California, Irvine, CA, USA

Institute of Computer Technology, TU Wien, Vienna, Austria
e-mail: amirr1@uci.edu

A.M. Rahmani et al. (eds.), *Fog Computing in the Internet of Things*,
DOI 10.1007/978-3-319-57639-8_8

devices with cloud-based services [10–12]. One of the main areas of focus in achieving this integration is the architecture enabling such an interoperability. A common approach is to directly connect the sensor devices to Cloud services. However, due to the resource constraints of the end-user devices, monitoring and actuating devices, an architecture with an intermediate computing layer is becoming the most widely used approach. This intermediate layer, known as Fog computing or edge computing [13–17], provides both generic and domain specific services for devices to enhance usability, reliability, performance, and scalability among others.

Several unique characteristics of healthcare demand the use of Fog computing in IoT-based health monitoring systems. First, the nature of the sensor devices (especially many wearable or implantable ones) require resource efficiency more than many other domains. Second, the nature of communication required by these sensors is often streaming type of transmission. For instance, electrocardiogram (ECG) signal collection requires a continuous communication with 4 kbps bandwidth per channel. Third, due to the criticality of the application domain, immediate response to important events gathered by sensor nodes is mandatory. The overall system needs a high level of reliability where patterns of physiological signals need to be recognized in real-time. Moreover, providing the freedom of movement of individual under a medical supervision through IoT devices is also a key requirement. These functions can be mainly supported by services in the Fog computing layer. Some of these services provided by the Fog layer are presented in this chapter focusing on the healthcare application domain. In general, this chapter concentrates on the following main areas:

- Healthcare IoT system requirements, along with services of the Fog computing layer to address them, are described.
- System architecture of a Fog-enabled healthcare IoT system is presented.
- Performance and advantages of the Fog computing layer services via a proof-of-concept full system implementation are demonstrated.

8.2 Healthcare Services in the Fog Layer

Fog computing layer provides computing, networking, storage, and other domain specific services for IoT systems. The healthcare domain has a set of requirements that uniquely identify it from other IoT applications; for instance, one of the use cases of healthcare IoT is remote monitoring, which demands a high degree of reliability. Unlike other domains, the security and privacy aspects of healthcare are also of critically important. This section highlights the services that can be provided by the Fog layer with a focus on healthcare. The physical proximity of Fog layer to Body Area Network (BAN) of sensors and actuators allows us to address the requirements of healthcare IoT. Some of these services are generic and can be used by any application domain. Figure 8.1 shows a generalized view of the services of the Fog layer, which are discussed separately in the following sections.

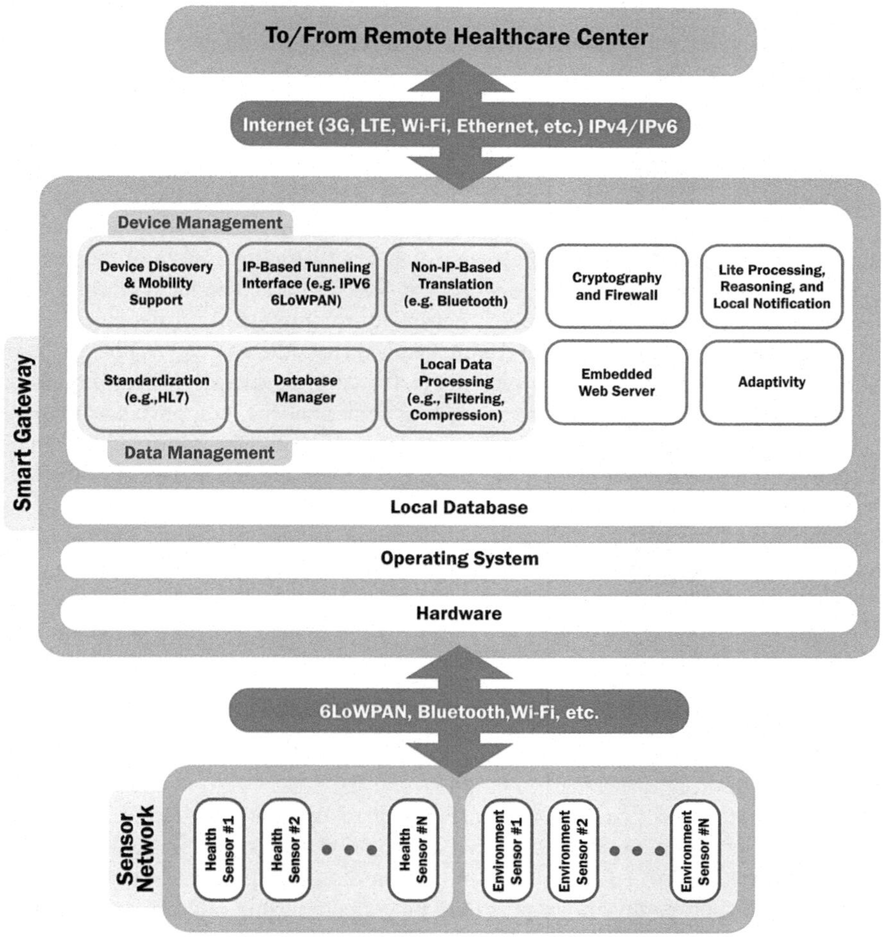

Fig. 8.1 Services provided by the Fog layer

8.2.1 *Data Management*

Data management has an important role in Fog computing by which the sensory data is locally processed to extract meaningful information for user feedback and notifications along with system plan adjustments. According to the system architecture, Fog layer continuously receives a large amount of sensory data in a short period of time from the sensor network, so it should manage the incoming data to provide a fast response regarding various user and system conditions. This task becomes more significant in healthcare scenarios since latency and uncertainty in decision making might cause irreversible damages for the patients. According to the different functionalities of the data management task, we introduce it in five different units, all of which are essential in a smart e-health gateway. These units are local storage, data filtering, data compression, data fusion, and data analysis.

8.2.1.1 Local Storage

The gateway needs to store received data from several sources in a local storage to be able to utilize it in the near future analysis. The operating system of the gateway has a file server and a database to store and recover data. The local storage in the gateway can be used to store files in encrypted or compressed format based on the type, size, and importance of data. Using local storage, the gateway is able to export data to medical standard formats such as Health Level Seven (HL7) [18] if required. Other functionalities of the gateway such as data analysis, compression, filtering, and encryption are also dependent on a temporary storage. The speed of data transfer between cloud layer and Fog layer is limited to network speed and most of the local calculations in the gateway are limited by its processing power. So, in case of any imbalance in computation time and transfer time, the local storage acts as a local cache memory to make data flow continuous. Local storage also helps saving data while the Internet connection between gateway and cloud server is not available. The gateways send saved data to the cloud when it connects to Internet again. This local storage is shown in Fig. 8.1 as database manager.

8.2.1.2 Data Filtering

Data filtering is the first data processing unit to implement filtering methods at the edge after receiving data from the sensor network. To obtain patient medical condition, various bio-signals such as electrocardiogram (ECG), electromyography (EMG), and photoplethysmogram (PPG) are collected using relevant probes. These signals usually include complex shapes with a small amplitude (i.e., millivolts) and consequently are susceptible to some unavoidable noises and distortions accumulated during the health monitoring. Such noises are thermal noise, electromagnetic interference noise, and electrode contact noise. Available light-weight filtering in some of the sensor nodes reduces these accumulated noises although it might be insufficient in practical cases. Therefore, the data filtering unit in the Fog layer enables to remove noise and to increase aspects of the signals (e.g., signal-to-noise ratio) using various filters (e.g., finite impulse response (FIR) filter) before any other local data analysis.

8.2.1.3 Data Compression

For reducing a large amount of transmitted data over a communication network, data can be compressed by lossless or lossy compression methods. In healthcare

IoT applications, lossless compression is more preferable in most of cases because lost data can cause inappropriate disease diagnosis. Although lossless compression algorithms can recover original data accurately, they are often complex. Correspondingly, they are not suitable for sensor nodes which are resource constrained regarding to battery capacity, computation, and memory capacity. For example, lossless ECG compression methods [19–21] cannot be run in many types of sensors. In some cases, these lossless algorithms can be successfully operated at sensor nodes but they cause a large power consumption and latency. Thanks to Fog computing, all mentioned limitations at sensors can be avoided by switching all burdens of sensor nodes to the gateway which handles these burdens efficiently while respecting the real-time requirement.

8.2.1.4 Data Fusion

Data fusion is the data processing unit to integrate sensory data from multiple sources to obtain more robust data and meaningful information. This processing unit efficiently decreased the volume of sensory data by removing redundant data and replacing new information. Consequently, this data reduction improves local data analysis and data transmission to the remote servers.

Data fusion can be divided into three classes as complementary, competitive, and cooperative [22]. First, complementary data fusion combines (at least) two different data from different sources to obtain a more comprehensive knowledge in the Fog layer. For instance, combination of patient health parameters with surrounding context data provides more data about patient condition. Second, competitive data fusion improves data quality as well as system decision making by integrating data collected from one source with (at least) two sensors. For example, a more robust heart rate value is extracted using values from an ECG signal and values from a respiration signal. Third, cooperative data fusion provides new information from one source using different sensors. Determination of the patient medical state using vital signs (e.g., heart rate and respiration rate) is an example of a cooperative data fusion.

8.2.1.5 Data Analysis

Data analysis unit at the edge enables the healthcare system to process the sensory data locally. This unit improves the system performance by decreasing response latency and data transmission to the cloud servers. For example, in case of patient health deterioration, the emergency event is detected and responded early and effectively since the data is processed locally instead of being transmitted to the cloud and awaited for the appropriate response.

In addition, the data analysis unit improves data reliability as well as system consistency. Connectivity losses and limited bandwidth for data transmission are unavoidable events in long-term remote health monitoring because the patient might engage various activities in different environments. Hence, data analysis at the edge could manage the system's functionality locally, store the sensory data as well as the calculations in a local storage, and subsequently synchronize the Fog layer with the remote cloud after reconnecting to the network.

8.2.2 Event Management

Several important events happen during patient monitoring which may be a change in vital signs, activities, or environment of the patient. Each event triggers a specific action in gateway or switches to a learned behavior in personalized systems. Fog computing provides low latency communication which helps to notify health experts, caregivers, and even patient very fast in case of a serious event. In such case when an immediate response is necessary in form of medical actions or automatic system actuation, the event management service ensures on time and proper signal delivery. The real-time and fast response of actuators are important in some medical events like changing the frequency of nerve stimulation according to heart rate or adjusting automatic insulin pump with blood glucose level. Other emergency events might also happen which required to notify rapid response team, caregivers, or family members of the patient.

8.2.3 Resource Efficiency

In healthcare IoT applications, resource efficiency is one of the most essential requirements because failure in resource management can cause serious consequences from sensor nodes' malfunction to imprecise disease diagnosis. Particularly, energy consumption of sensor nodes and latency of gathered data presented at end-users' terminals must be attentively considered. They are discussed in detail in the following.

8.2.3.1 Energy Efficiency of Nodes

Sensor nodes in health monitoring systems are typically small and resource constrained, such as small battery capacity, but it is required that sensor nodes must be able to operate in an appropriate duration such as a whole day or even a few days. In order to fulfill these requirements, sensor nodes must operate efficiently in terms of energy consumption. Several methods including software and hardware-based techniques can be applied for achieving the task. For instance, sensor node's hardware should be designed for particular purposes instead of general tasks. This method helps to save energy consumption by avoiding unused components or high power consuming components. However, it is more challenging for designing an energy efficient nodes than customizing software running at the nodes. Particularly, the software must be able to perform primary tasks sensor nodes while it must be extremely simple for reducing computation time. For example, sensor nodes in IoT-based fall detection systems only gather digital data and transmit the data to Fog layer which runs complex fall detection algorithms [23]. Accordingly, sensor nodes' power consumption can be reduced dramatically. In another example, several approaches [24, 25] show high levels of energy efficiency at sensor nodes when running ECG feature extractions at Fog layer instead of sensor nodes.

8.2.3.2 Latency

In health monitoring applications, latency is critical because it can cause inappropriate disease analysis and delay in decision making. Latency in IoT-based application comes from both processing and transmitting of data such as transmission latency from sensor nodes to end-users via gateways, Cloud, and processing at sensor nodes, gateways. In many cases, transmission latency and processing latency are often in a trade-off relationship. However, processing data does not always guarantee to reduce the total latency. In some cases, data processing even increases the total latency. In order to reduce the total latency and fulfill time requirements of real-time health monitoring, a distinct method must be applied for a particular sensor nodes and applications. In other cases, simple filtering methods for eliminating noise and invalid data can help to reduce a large amount of data transmission as well as the total latency. For example, in an ECG feature extraction application, the feature extraction algorithm must be run at Fog instead of sensor nodes for reducing the total latency[25].

8.2.4 Device Management

Device management encompasses many areas of IoT infrastructure. In this section, the focus of the discussion is on device management from the point of view of device discovery and maintaining connectivity during mobility.

8.2.4.1 Discovery and Mobility

The resource constraints of devices in the sensor and actuator network have been mentioned briefly earlier. For battery powered devices, the lifetime of the battery is precious and needs a proper management. Devices should go to sleep state in a controlled way whenever they become idle. Any communication that occurs when the device is in the sleep state needs to be taken care of by the Fog layer. In healthcare scene, a patient wearing medical sensors and moving from a location to another changes the corresponding gateway that handles communication.

This means that a sleeping sensor can wake up at a vicinity of different gateway, i.e., it was connected to another gateway when it entered to the sleep mode. Device discovery service helps another device that requires connection to a sleeping sensor node to discover it and gracefully handle the sleeping cycle. A detailed function of this service can be found at [26].

To provide the freedom of mobility for the patient, and to do so without consuming much of the scarce resources, the Fog layer service is used to handover the locally retained information from one gateway to another. Once the device is discovered in a new location, the handover takes place to seamlessly continue the monitoring process at the new place. A simplified implementation of mobility and device discovery working together is presented in [26]. These services, like some of the other Fog computing services, can also be used for other domains as well.

8.2.4.2 Interoperability

The Internet of Things is composed of heterogeneous set of communication protocols, platforms, and data formats. There are many standardization efforts to establish uniformity among the different components. The current vertical segmentation of applications need to be bridged to ease the creation of holistic healthcare applications. Traditionally, interoperability is challenged due to the resource constraints in the majority of end-devices. The Fog computing layer plays a significant role in providing services that leverage the proximity of this layer to end devices and hence ease interoperability. These services act as an adapter among different communication protocols, data formats, and platforms. Preliminary work can also be accomplished at the Fog layer to provide semantic interoperability at the cloud to give meaning to the data collected through the sensors.

8.2.5 Personalization

System behavior can be configured for different applications of Fog computing in advance or at the run time. This, however, might be insufficient for healthcare scenarios since users might have various medical conditions and engage different activates in different environments. Therefore, a dynamic plan for the system is required to not only personalize the system behavior according to the user requirements but also adaptively adjust the system over time, especially in emergency situations. In this regard, it improves the health applications (e.g., local decision making) as well as optimizing the system performance (e.g., energy efficiency).

Personalized system behavior can be defined for various health applications using rule-based techniques and machine learning algorithms. To this end, different priorities and modes are defined for the system parameters (e.g., sensor sampling rate and data transmission rate), and according to the patient conditions, the appropriate values are selected. In addition, the priorities become personalized regarding the medical history of the patient. In a simple example, if a heart failure was detected for a patient during the monitoring, the system would learn to increase the priorities to heart related parameters.

8.2.6 Privacy and Security

In general, security is critical for all applications and it is more essential in cases of healthcare because a single insecure point in a system might cost a human life. For example, it is reported that a insulin pumper in a IoT glucose management system can be hacked within 100 feet [27]. In order to provide a secure IoT healthcare system, the whole system including sensor nodes, gateways, Fog and Cloud must be attentively considered. If one of the devices or components is hacked, the entire system can be controlled or manipulated by hackers. For example, several methods such as AES-128 or CMAC-AES-128 can be applied at sensor nodes and gateways for data encryption and decryption, respectively. At gateways, IPtable offered by Linux can be used for configuring IP tables giving grant permissions to particular communication ports [28]. Although these methods can improve some security levels, it cannot be seen as robust methods for protecting the entire system. On the other hand, other approaches in the literature [29–31] provide a high level of security. However, they cannot be applied in IoT systems because their complex cryptographic algorithms are not suitable to resource-constrained sensor nodes. To address the security issues in IoT healthcare systems, Rahimi et al. [32, 33] recently introduce end-to-end secure framework. The framework can provide efficient authentication and authorization for Health-IoT systems while the main parts, consisting of several complex security algorithms, are run at Fog layer.

8.3 System Architecture of Healthcare IoT

The architecture of a system provides information about the components, interactions, and the organization of the parts. It is one of the key elements for achieving graceful scaling and performance. Moreover, it is designed to meet the functional requirements of the application domain. Among the non-functional requirements that constrain the system architecture design, few of these are scalability, usability, and performance. One of the main challenges is the huge number of devices that are getting connected to the Internet. Connecting more devices cause the available resources, such as bandwidth and computing power, to be shared by more nodes leading to quality and performance degradation. However, the criticality of the application domain makes this degraded infrastructure unacceptable. In addition, a large proportion of these devices are resource constrained. This shortage of resources add more design constraints to the architecture design.

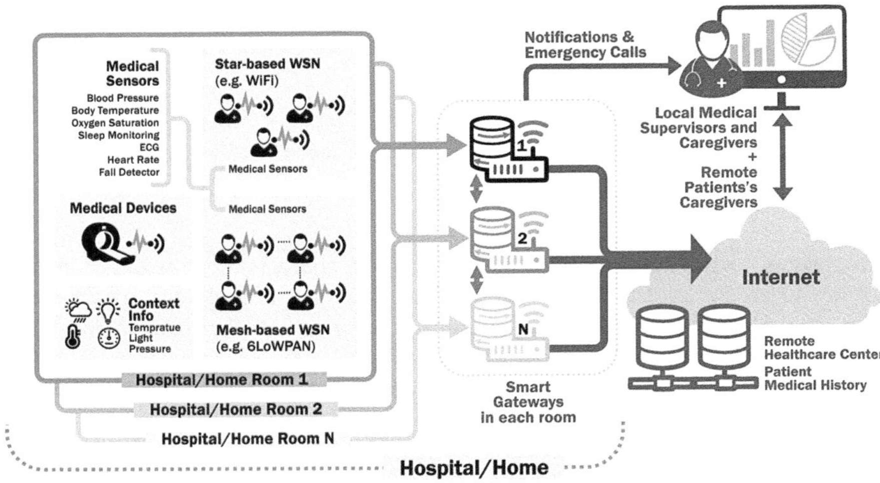

Fig. 8.2 Healthcare IoT system architecture

The introduction of Fog computing layer between the end devices and the cloud has contributed to a design shift towards IoT-based systems thereby alleviating the challenges mentioned above. The Fog layer can be used to provide a wide variety of services to support the resource-constrained nodes [13]. In a healthcare IoT application, an overview of the architecture is shown in Fig. 8.2, where resource-constrained nodes can be wearable or implantable sensors and actuators. A patient can be at home or in hospital and the sensors send the values of the physiological signals they read to a local gateway in the Fog layer. The Fog layer has local services handling data, events, devices, and the network. In healthcare scenario, this architecture is composed of the following three main parts in each layer.

1. **Medical sensors and actuators:** Mainly connected through low power wireless communication protocols, serves in identifying subject, reading physiological signals and acts in response to a command from the Fog layer.
2. **Smart e-health gateways:** Distributed network of gateways form the Fog layer and serve the underlying sensor and actuator network. It usually has multiple interfaces that enable it to communicate with different protocols. It hosts a wide variety of services and acts as a bridge to the cloud. This is the main focus of the chapter.
3. **Cloud platform:** Back-end system where the data persists and stakeholders get access to the system, through web or mobile interface. It is the point of interface with other systems, such as hospital information system and patient health records.

8.4 Case Study, Experiments, and Evaluation

Previous sections discussed theoretical aspects of Fog computing and highlight its benefits in healthcare IoT systems. In this section, we discuss the detailed implementations and experiments conducted in our work (Fig. 8.3).

The architecture of our system consists of three layers, shown in Fig. 8.4. The first layer is sensor network containing three group of sensors: a set of medical sensors to record vital signs (heart rate, respiration rate, body temperature, blood pressure, blood oxygen level, and ECG), a set of environment sensors to find the situation of the patient (light, temperature, and humidity), and activity sensors to find the body movements, posture, and step count.

The Fog computing takes place in the second layer where a network of gateways collect data from sensor nodes via Bluetooth and Wi-Fi wireless communication. A Bluetooth service and Node.js UDP server receive and store the data in separate files for each sensor and each patient. An Apache server also runs a service in the background which calls a Python script to read and process data from local files. Because data is collected from several sources with different properties, the gateway performs an adaptation to the data. The adaptation includes handling different

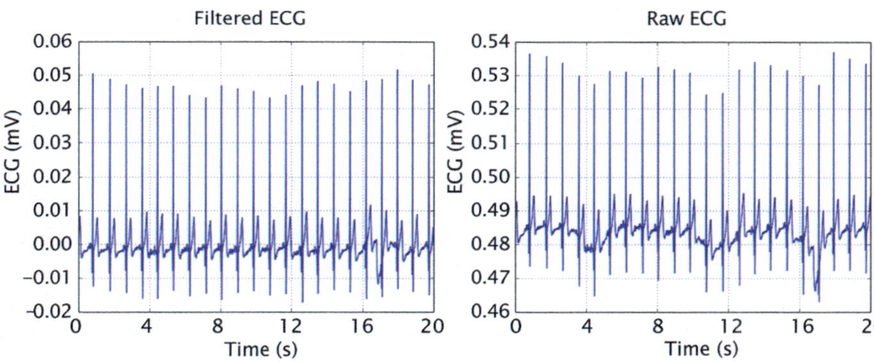

Fig. 8.3 Raw and filtered ECG data

Fig. 8.4 Healthcare IoT system architecture

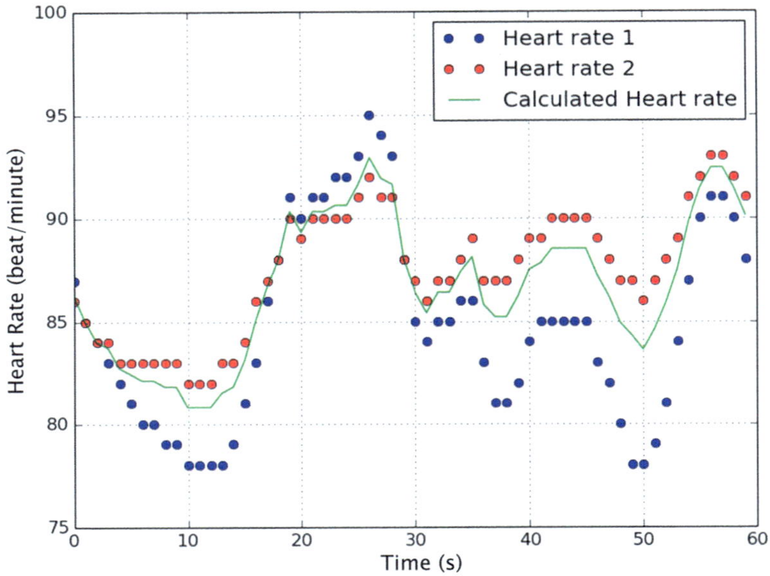

Fig. 8.5 Heart rate data fusion

communication protocols (in UDP server script) and unifying the sampling rate which varies from five samples per day for step count to 250 samples per second for ECG (via Python script). The python script is also responsible for preliminary data analysis to detect abnormalities before in-depth analysis at the cloud layer. This service first reduces the signal noise from ECG signal using a bandpass filter (0.5–100 Hz) with finite impulse response (FIR). Then using RR intervals the heart rate is calculated. The raw and filtered ECG signal are demonstrated in Fig. 8.3.

Data fusion is implemented on the sensory data in the Fog layer. In this case, two heart rate signals are acquired from the user using two different devices. As illustrated in Fig. 8.5 with blue and red dots, the heart rate values from two devices may not be identical due to the noise and measurement inaccuracy. In our approach, first, we remove outliers and out-of-range heart rate values (e.g., zero values) from the data. These outliers are mainly because of loose probe connections over the course of monitoring. Second, we implement a weighted average on the two heart rate values to improve the signal-to-noise ratio. The green line in Fig. 8.5 shows the calculated heart rate. In the calculation, the average weights are defined with respect to the sensor accuracy mentioned in data-sheets.

In addition to the aforementioned data processing, data compression is implemented in the Fog layer. Before transmitting data to a remote cloud, the data is compressed and stored in the local storage of the Fog layer to provide a backup in case of Internet connectivity is lost. In this case study, *tar* method is utilized to create a temporary file for the sensory data, and then the file is compressed using *tar.gz* method while the size of the file is increased to a certain value. We set this value to 500 KBytes. The relation between the compression ratio and the file size is represented in Table 8.4. As illustrated, in larger file size, the compression ratio is higher although this improvement is insignificant for the files larger than 500 KByte.

In order to improve the data security, an asymmetric encryption method using *Crypto* library [34] in Python is implemented in the Fog layer. With this method, the compressed data is encrypted with a public key in the Fog layer and is decrypted with a private key by the data collector service in the cloud.

There is also a local storage service in the Fog layer which consists of a file server to store the files and a MySQL database to keep the indexes and attributes of them. One objective of this case study is to store data in the cloud and use the benefit of available processing power of the server.

The gateway in the Fog layer transmits collected data to the cloud server when the Internet connection is available. When the Internet is unavailable, the gateway marks unsent files in the database and tries to re-transmit once the Internet connection is re-established. The local storage service periodically synchronizes the database with the cloud server and removes out-dated and duplicated files from the gateway storage and database.

The third layer in the architecture of our system is the cloud server which is responsible for receiving data from the Fog layer, processing, and storing. Cloud server uses stored data together with the history of the patient to analyze the health status of a patient. An interface for caregivers is created to send alerts, reports and plot vital signs and other sensory data in real-time. Figure 8.6 shows the web-based user interface developed by HTML5 WebSocket. The cloud server provides also information and notifications via a mobile application for patients and caregivers.

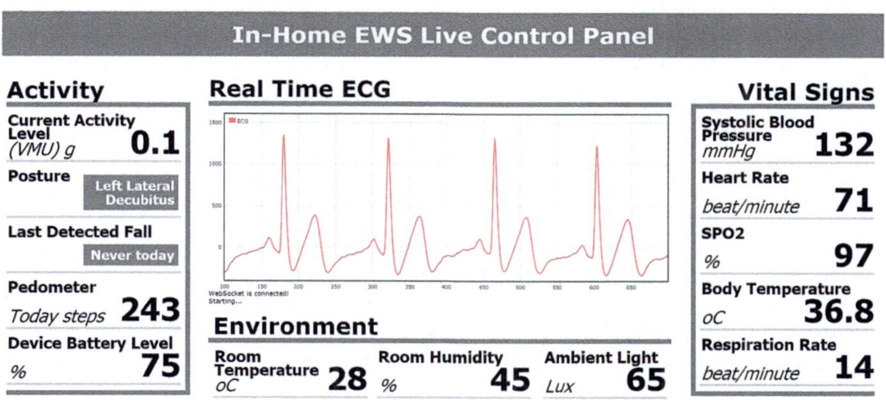

Fig. 8.6 Live control panel web interface

8.4.1 Performance Measures

The entire system including sensor nodes, smart gateways with Fog, Cloud, and client back-end parts is implemented. The sensor nodes are able to gather bio-signals, i.e., ECG, EMG and contextual data, i.e., temperature and humidity. The sensor nodes are created by a combination of medical sensors, micro-controller, and wireless communication IC. For evaluating interoperability of our gateways, several types of sensor nodes are used such as Bluetooth nodes, Wi-Fi nodes, and 6LoWPAN nodes. These nodes are formed into a mesh or star network depending on particular wireless communication protocols. In detail, Bluetooth nodes, including Bluetooth Classic and Bluetooth low energy (BLE), are created by integrating low-cost HC-05 and BLE ICs into micro-controllers (i.e., ATMEGA128, ATMEGA328P, ARM Cortex M3), respectively. Similarly, Wi-Fi nodes are constructed by using the same micro-controllers and low-cost ESP8266 boards supporting a high speed Wi-Fi communication while the 6LoWPAN node is built from a CC2538 module. For gathering data, these nodes are connected with several sensors via SPI, I2C, or UART. For example, in order to collect EEG and EMG signals, these devices are connected with analog front end devices (i.e., TI ADS1292 [35] and TI ADS1298 [36]). With the purpose of efficient managing resources of sensor nodes, i.e., power consumption and hardware distribution, embedded operating systems are installed in sensor nodes. For instance, RTX, FreeRTOS, and Contiki are used in Wi-Fi, Bluetooth, and 6LoWPAN nodes, respectively [37–39]. Specification and power consumption of these sensor nodes when transmitting data at 8.7 kbps are presented in Tables 8.1 and 8.2.

Gateway of the system is built from several devices such as Pandaboard [40] and Texas Instruments (TI) SmartRF06 board integrated with CC2538 module [41] and MOD-ENC28J60 Ethernet Module [42]. Due to the Pandaboard's advantages of high performance 1.2 Ghz dual ARM Cortex-A9 core processors, large amount of memory and hard disk capability, the Pandaboard is used as a core component of the gateway for performing algorithms and services [40]. In addition, Pandaboard can support several operating systems comprising of Windows CE, WinMobile,

Table 8.1 Hardware specification of sensor nodes and UT-GATE

Device	Micro-controller	Flash (KB)	RAM (KB)	EEPROM (KB)	Clock (MHz)	Voltage (V)
Zigduino R2	ATMega 128	128	16	4	16	3.3
Arduino Uno R3	ATMega 328P	32	2	1	16	5
Arduino Mega	ATMega 1280	128	8	4	16	5
Arduino Due	ARM CortexM3	512	86	–	84	3.3
Zolertia Z1	MSP430	92	8	–	16	3.3
TI-CC2538	ARM Cortex M3	Up to 512	32	–	32	3.3
Pandaboard	Dual-core ARM Cortex-A9	Up to 32,000	1000	–	1200	5

Table 8.2 Power consumption of sensor nodes when transmitting at 8.7 kbps

Communication type	Current (mA)	Voltage (V)	Power consumption (mW)
6LoWPAN node	24.6	3.3	81.2
Wi-Fi node	114	3.3	376.2
Bluetooth 2.0 node	56.9	3.3	187.7
BLE node	31.6	3.3	104.4

Table 8.3 Sensing to actuation latency comparison for local Fog computing-based vs. remote cloud-based scenarios

Latency of the sensing-to-actuation loop	Using Wi-Fi (ms)	Using BLE (ms)
Fog-based (locally via UT-GATE)	21	33
Cloud-based (remotely via the Cloud)	161	176

Symbian, Linux, and Palm. By applying one of the operating systems, resources can be managed efficiently. For example, collision caused by simultaneous access to the same hardware can be avoided by hardware abstraction in Windows and Linux. Although the Pandaboard cannot support all wireless communication protocols, it supports popular wireless and wire communication protocols such as 802.11 b/g/n, Bluetooth, and Ethernet. For dealing with other wireless protocols, the Pandaboard is equipped with Texas Instruments (TI) SmartRF06 board which is integrated with CC2538 module for supporting Zigbee and 6LoWPAN [41].

Although Fog computing is capable of providing a large number of advanced services such as local storage and push notification, roles of Cloud cannot be lessened. For example, limitations of the local storage of Fog computing units, such as a limited storage capacity and limited accessibility, can be solved by Cloud connection. In our implementation, the remote server is built in Cloud for handling client requests, running complex algorithms, enhancing Fog layer services, and responding with data by providing the requested data along with graphical user interface. The free service provided by "heliohost.org" including MySQL server with remote access facility is used.

Depending on particular services, Fog or Cloud or a combination of Fog and Cloud computing should be used to implement the services. For example, Table 8.3 shows that decision making at Fog layer is more efficient in terms of latency than at Cloud. Two protocols including Wi-Fi and BLE are used during latency measurements. The currently implemented functionalities of the gateway, called UT-GATE, such as data fusion, data compression, local storage are discussed in detail in the following subsections.

8.4.2 Data Compression at Fog Layer

To reduce the amount of transmitted data, and hence to make the system more energy efficient, data compression is applied at the Fog layer. The compression rate depends on a particular data compression method. For example, lossy compression methods have a high compression ratio such as 50:1 while lossless compression method can achieve compression ratio about 8:1–9:1. As mentioned, loss of data can cause serious consequences in a healthcare domain. Therefore, a lossless compression is applied in our application for compressing bio-signals such as ECG and EMG. However, not all lossless compression methods can fulfill the latency requirements of real-time applications defined by IEEE 1073 [43]. In our implementation, an LZW [44] algorithm is used at Fog layer because it provides rapid compression and decompression without losing any data. Computation efficiency of the LZW compression algorithm increases when the number of inputs increases. For example, computation time increases 8 times when the input size increases 10 times. However, in this case, latency raises dramatically. Hence, it is recommended that the data size of inputs used for data compression must be carefully chosen to achieve the computational efficiency and fulfill the real-time requirements of health monitoring.

Table 8.4 shows time results of LZW compression and decompression at the gateway. In our application, EMG data is acquired from eight channels with a data rate of 1000 samples/second per channel in which each sample size is 24bit. The LZW algorithm uses 120 samples as its inputs for achieving a high computational efficiency. The number of sensor nodes connected to the gateway varies from 1 to 50 nodes. The results show that when the gateway receives more data, it operates more efficiently.

Table 8.4 Compression results at UT-GATE and latency reduction

Number of sensor nodes connected to UT-GATE	1	2	5	10	50
Number of analog channels	8	8	8	8	8
Data size (120 samples) (B)	8400	16,800	42,000	84,000	420,000
Compressed data size (B)	808	1597	3893	7696	38,333
Compression time (ms)	3.1	4.4	9.2	16.6	73.0
Decompression time (ms)	3.3	4.6	11.3	23.0	83.9
Total time of comp., tran., and decomp. (ms)	12.86	21.77	51.64	101.16	463.5
Transmission time of non-processed data (ms)	67.2	134.4	336	672	3360
Total latency reduction (%)	80.8	83.8	84.6	84.8	86.1

8.4.3 Benefits of Data Processing at the Fog Layer

In real-time health monitoring applications, real-time streaming data is captured and processed for health indicators. In an IoT-based monitoring system shown in Fig. 8.2, computation resources for signal processing are distributed in each layer but with different capacities. As illustrated in Sect. 8.2.3, the benefits of processing streaming data in Fog layer on the system are twofold:

- Bring energy efficiency to sensor nodes by shifting streaming data processing from sensor node to Fog layer;
- Shorten data transmission latency from sensor nodes to cloud by reducing data size.

To demonstrate these two benefits, tests in three scenarios are conducted in the implemented system. They are, applying signal processing on (1) a sensor node, where gateway receives and passes processed data to cloud, (2) gateway, where raw data is received and processed data is transmitted to cloud, which can be also considered as Fog-assisted cloud computing, and (3) the cloud, where raw data is directly passed to it through sensor node and Fog layers. The algorithm of ECG signal processing for *Q*, *R*, and *T* wave extraction is applied on MIT-BIT Arrhythmia database [45]. The signal processing on ECG data includes noise reduction, wavelet decomposition, and peak detection for extracting waves, as shown in Fig. 8.7.

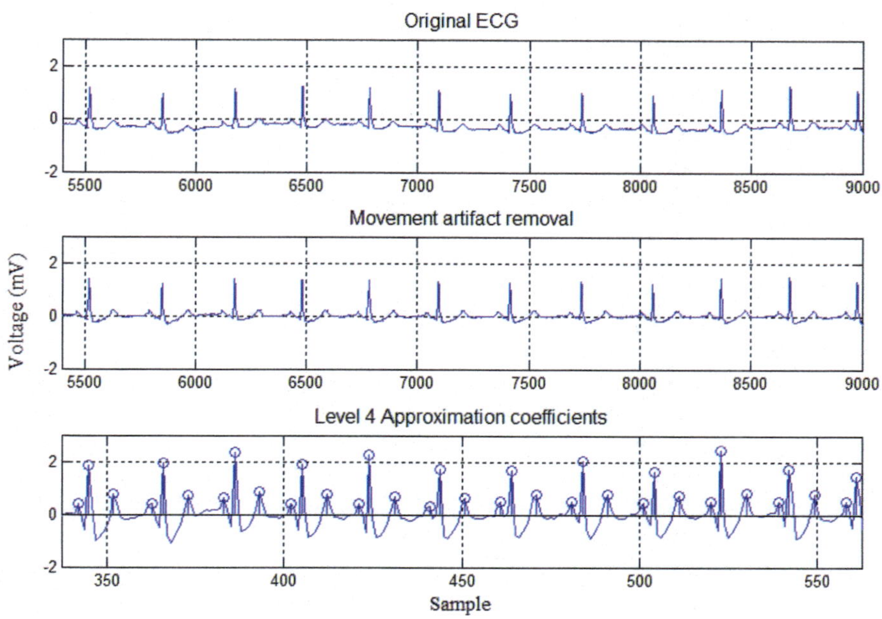

Fig. 8.7 ECG processing implementation

Table 8.5 Providing energy efficiency for sensor nodes

	Time	Current	Energy
Processing at sensor node (collection + execution)	2.78 s + 101 ms	10.1 mA + 106.8 mA	127.34 mJ
Process at gateway/cloud (transmitting raw data)	95 ms	180 mA	56.34 mJ

Table 8.6 Number of transmitted samples from Fog to cloud

	Processing at sensor node or UT-GATE	Processing at cloud	Improvement (%)
No. of samples	259	1000	74.1

Table 8.7 Latency reduction between gateways and cloud server for 240 KB of raw samples

Network condition	Data rate (Mbits/s)	Raw samples trans. time (ms)	Raw samples proc. time + trans. time of processed samples (ms)	Latency reduction (%)
Light load	18	106.6	96.3 + 6.6	3.5
Medium load	12	152.2	96.3 + 9.5	30.5
Heavy load	9	213.3	96.3 + 13.5	48.5

In addition to peak detection, heart rate is calculated from *R* to *R* interval. In these three scenarios, energy consumption of sensor node, sample size delivered from gateway to cloud, and its data transmitting time are measured and calculated for comparison.

Regarding energy efficiency test on sensor nodes, 1000 samples of ECG data (360 samples per second) are saved at Aruidno Due, which together with ESP8266 Wi-Fi module acting as a sensor node. To simulate real-time ECG data processing on sensor node for every 1000 samples, the sensor node first waits for data gathering and then proceed with processing and transmission. The overheads of signal processing on sensor node regarding energy consumption are accessed by monitoring execution time and current consumption. The results from three scenarios can be seen in Table 8.5, where about 55.7% can be saved for sensor node when the outsourcing signal processing task to posterior layers.

To look into the benefits of Fog computing on data transmission latency between gateway and cloud server, sample size is first calculated in three scenarios, presented in Table 8.6. The saved data transmission time from Fog to cloud in practice under three different network conditions is presented in Table 8.7. It can be seen that, for every 1000 ECG data samples, sample size is reduced by 74.1% at cloud due to signal processing at two prior layers. Both signal processing time in Fog layer and data transmission time are considered as transmission latency. As shown in Table 8.7, transmission latency is evidently reduced especially in Wi-Fi network with heavy load.

Table 8.8 XML Status code and description

Code	Description
0	Invalid request or Error
1	Notification – {total in number}
2	No new notification

8.4.4 Local Storage, Notification, and Security at the Fog Layer

The majority of the data management Fog services explained in Sect. 8.2.1 make use of the local storage function of the gateway in the Fog layer. Incoming data from 6LoWPAN modules through a UDP server in the gateway, on port 5700, and from RTX Wi-Fi modules is aggregated in the storage. The Bluetooth module on the gateway also receives data from Bluetooth sensor modules. The local storage is implemented using MySQL database, to leverage the federation feature provided by the database engine for data synchronization with the cloud storage. The same table schema is used both in the cloud and in the gateways allowing easy migration of the data without mapping to other format.

The data stored locally is stored for about 30 min while the synchronization takes place. In the condition that there is a failure in connectivity with the cloud, the gateway stores the data for as long as there is connectivity or it runs low on memory. In the case that the connectivity cannot be maintained, the gateway removes the old data and also provides additional user services, such as notification and access to the local data. The notification service can be used as mentioned above or can also be configured to run along with the cloud-based notification service.

For protecting the system, an end-to-end secure schema is implemented. The schema provide a high level of authentication and authorization for end-users. In addition, the schema guarantee that the system is secure whilst sensor nodes do not need to be reconfigured in cases of sensor node mobility. The detailed information and an implementation of the schema are presented in [33].

Notification service in the gateway can be immediately triggered when receiving an alert signal from other services such as ECG feature extraction or an early warning service. In order to reduce the burden of Fog layer and provide a global notification, the notification service is primary built at Cloud. The push notification at Fog layer is merely responsible for notifying the push notification in Cloud by sending signals in XML formats, shown in Table 8.8. In addition to the push notification service at Fog and Cloud, a mobile application is created for supporting the push notification service. When it receives the incoming signal from the notification service at Cloud, it immediately pops up a text message on an end-user's phone to inform about an alert case.

8.4.5 WebSocket Server for Interoperability

To enhance the interoperability of the healthcare IoT implementation, we implemented an embedded websocket server at the Fog layer. It is written using a Python framework, known as Tornado, which is an asynchronous web server framework. The server listens to UDP connections and a full duplex communication can be established with a client node. This setup is used to collect ECG signals coming from sensor nodes. Each message contains 800 bytes of data from 400 samples of ECG, at an average speed of 1.1 KB/s. Client nodes with an interface can also access an HTML page hosted on the gateway to see the received information rendered into a graph. This can be extended to enhance the service to listen to various transport layer protocol sockets to enable interoperability.

8.5 Related Applications of Fog Computing

Three layers of physical separation in IoT architectures is a common approach in recent years. However, prior to that, a client server approach between sensor nodes and the cloud was a usual implementation. In both cases, amalgamation of the cloud computing and IoT technologies provides significant benefits in the healthcare application domain [46]. The IoT allows integrating objects into an information network, so in health scenarios, different medical sensors could collect health data from the patient continuously. The data is transmitted to a remote cloud server for analysis and decision making concerning requirements of the use cases. Using data analytics and machine learning approaches, patient medical condition is estimated, decision making is fulfilled, and notifications are provided for the patient and health professionals [47, 48]. In another work, Bimschas et al. [49] implement basic learning and intelligence in the gateways of the Fog layer through application code execution to allow them convert between protocols, smart caching, and discovery.

The application of Fog computing to different domains is also an active research area. Related to this chapter of the application of Fog computing to healthcare, Shi et al. [50] have used it in a similar context. Their work motivates the need for Fog computing and highlights the low-latency requirements of healthcare applications. In addition, it also highlights the resource constraints of sensor and actuators at the end of the IoT network. To meet the latency requirement and support the end devices, their implementation of Fog computing provides services, such as online data analytics and interoperability of various communication protocols. In another work [51] that applies Fog computing for healthcare, they have developed a gateway to be used in the Fog layer to provide services for healthcare domain. Similar to the previous work, the gateway provides interoperability to various network protocols of sensor nodes and communicates with the cloud through Wi-Fi or Ethernet. The gateway is used for a medical application, emergency attendance in a patient care facility. It is also used in monitoring patients to follow up on their medication. Moreover, their work considers semantic interoperability of the data for better quality of the medical information collected. In contrast, this chapter begins with more comprehensive general Fog computing services that enable healthcare applications and show specific use cases that consume these services. Moreover, Stantchev et al. in their work [52] present the advantages of applying the three tiered approach for healthcare scenario. It focuses on servitization and business aspect of healthcare IoT. In contrast to the work presented in this chapter, their work focuses on high level architectural aspects while this chapter presents a real world case study and experimental evaluation of the services.

Leveraging Fog computing in a healthcare applications can enhance latency of the system and energy efficiency to sensor node devices, as discussed in this chapter. These benefits of Fog computing are particularly important to healthcare applications which are sensitive to response time and need distributed analytics. In addition to the case studies of early warning score and ECG feature extraction, some other remote healthcare monitoring applications have been proposed to employ Fog-assisted architecture. Cao et al. [53] implemented a fall monitor application for stroke mitigation in an architecture with Fog computing with minimal response time and energy consumption. Moneiro et al. [54] showed the efficacy of Fog computing in their speech tele-treatment application, where speech data collected from smartwatch is stored and processed in Fog layer and only speech features are sent to secure cloud. Similarly, applications including pervasive brain monitoring with EEG, emergency response system integrating heterogeneous data, and ambient assisted living for seniors have considered to utilize system structure with Fog [55–57]. Moreover, as mentioned above, security plays an important role in all health monitoring IoT-based systems. In [33], authors provide an end-to-end Fog-based schema for enhancing security levels of health monitoring IoT-based system. Their results show that by applying the secure schema in the Fog layer, the entire IoT system can be protected even in case of sensor nodes' mobility. In [25], the authors show that latency and a large amount of transmitted data can be decreased dramatically by applying the services of the Fog layer.

8.6 Conclusions

This chapter focused on applying Fog computing to healthcare Internet of Things domain. As a case study, to highlight the benefit of Fog computing, a set of services was presented that enable healthcare IoT and utilized in an implementation of a smart gateway for Fog computing. These services are designed to address the key challenges in IoT, with a focus on addressing healthcare functional requirements. A geographically distributed network of these smart gateways, each handling a group of sensor nodes or patients, form the Fog layer in this architecture. This cluster of gateways provide a continuous patient monitoring means without limiting the movement of the patient in the coverage area.

Fog computing provides services for remote patient monitoring by reducing communication latency and improving system consistency. To this end, the patient's vital signs were processed locally, and a local notification was provided for the user. Moreover, for further analysis, the sensory data along with the obtained results were transmitted to the cloud server. The details of the Fog services and the obtained benefits were analyzed and performance measures presented. Some services can be duplicated or partitioned to multiple, such as sensor layer with the Fog layer or the Fog layer with the Cloud. This implementation specifics depend on the particular services and the functionality. Incorrect decisions of implementation location can cause inefficiency in terms of energy consumption, latency, and performance. For example, before sending data from sensor nodes to Fog layer, noise should be initially removed at nodes. Fog can be used to implement advanced and complex noise elimination and signal processing methods for enhancing the quality of collected data. The cloud layer should be used for the implementation of more advanced services, such as complex machine learning algorithms. In order to achieve a high level of energy efficiency at sensor nodes, processing and communication must be taken into consideration. Finally, specific medical use cases can have additional design constraints that demand fine tuning of behaviors of the Fog layer services.

References

1. European Commission Information Society, Internet of Things Strategic Research Roadmap (2009), http://www.internet-of-things-research.eu/. Accessed 14 July 2015
2. European Commission Information Society, Internet of Things in 2020: a Roadmap for the Future (2008), http://www.iot-visitthefuture.eu. Accessed 14 July 2015
3. A. Dohr, R. Modre-Opsrian, M. Drobics, D. Hayn, G. Schreier, The internet of things for ambient assisted living, in *Proceedings of the International Conference on Information Technology: New Generations* (2010), pp. 804–809
4. D. Miorandi, S. Sicari, F. De Pellegrini, I. Chlamtac, Internet of things: vision, applications and research challenges. Ad Hoc Networks **10**(7), 1497–1516 (2012)
5. M. Carmen Domingo, An overview of the internet of things for people with disabilities. J. Netw. Comput. Appl. **35**(2), 584–596 (2012)
6. H. Yan, L. Da Xu, Z. Bi, Z. Pang, J. Zhang, Y. Chen, An emerging technology – wearable wireless sensor networks with applications in human health condition monitoring. J. Manage. Anal. **2**(2), 121–137 (2015)
7. Y.J. Fan, Y.H. Yin, L.D. Xu, Y. Zeng, F. Wu, Iot-based smart rehabilitation system. IEEE Trans. Ind. Inf. **10**(2), 1568–1577 (2014)
8. B. Farahani, F. Firouzi, V. Chang, M. Badaroglu, K. Mankodiya, Towards fog-driven IoT eHealth: promises and challenges of IoT in medicine and healthcare. Elsevier Future Generation Computer Systems (2017)
9. C.E. Koop, R. Mosher, L. Kun, J. Geiling, E. Grigg, S. Long, C. Macedonia, R. Merrell, R. Satava, J. Rosen, Future delivery of health care: Cybercare. IEEE Eng. Med. Biol. Mag. **27**(6), 29–38 (2008)
10. European Research Cluster on the Internet of Things, IoT semantic interoperability: research challenges, best practices, solutions and next steps, in *IERC AC4 Manifesto - "Present and Future"* (2014)

11. B. Xu, L.D. Xu, H. Cai, C. Xie, J. Hu, F. Bu, Ubiquitous data accessing method in IoT-based information system for emergency medical services. IEEE Trans. Ind. Inf. **10**(2), 1578–1586 (2014)
12. L. Jiang, L.D. Xu, H. Cai, Z. Jiang, F. Bu, B. Xu, An IoT-oriented data storage framework in cloud computing platform. IEEE Trans. Ind. Inf. **10**(2), 1443–1451 (2014)
13. F. Bonomi, R. Milito, J. Zhu, S. Addepalli, Fog computing and its role in the internet of things, in *Proceedings of the First Edition of the MCC Workshop on Mobile Cloud Computing* (2012), pp. 13–16
14. M. Aazam, E.N. Huh, Fog computing micro datacenter based dynamic resource estimation and pricing model for IoT, in *2015 IEEE 29th International Conference on Advanced Information Networking and Applications* (2015), pp. 687–694
15. M. Aazam, E.N. Huh, Fog computing and smart gateway based communication for cloud of things, in *2014 International Conference on Future Internet of Things and Cloud (FiCloud)* (2014), pp. 464–470
16. A.-M. Rahmani, N.K. Thanigaivelan, T.N. Gia, J. Granados, B. Negash, P. Liljeberg, H. Tenhunen, Smart e-Health gateway: bringing intelligence to IoT-based ubiquitous healthcare systems, in *Proceeding of 12th Annual IEEE Consumer Communications and Networking Conference* (2015), pp. 826–834
17. A.M. Rahmani, T.N. Gia, B. Negash, A. Anzanpour, I. Azimi, M. Jiang, P. Liljeberg, Exploiting smart e-health gateways at the edge of healthcare internet-of-things: a fog computing approach. Futur. Gener. Comput. Syst. (2017). In Press, Corrected Proof. https://doi.org/10.1016/j.future.2017.02.014
18. Health Level Seven Int'l, Introduction to HL7 Standards (2012). www.hl7.org/implement/standards. Accessed 30 July 2015
19. M.L. Hilton, Wavelet and wavelet packet compression of electrocardiograms. IEEE Trans. Biomed. **44**(5), 394–402 (1997)
20. Z. Lu, D. Youn Kim, W.A. Pearlman, Wavelet compression of ECG signals by the set partitioning in hierarchical trees algorithm. IEEE Trans. Biomed. **47**(7), 849–856 (2000)
21. R. Benzid, A. Messaoudi, A. Boussaad, Constrained ECG compression algorithm using the block-based discrete cosine transform. Digital Signal Process. **18**(1), 56–64 (2008)
22. H.F. Durrant-Whyte, Sensor models and multisensor integration. Int. J. Rob. Res. **7**(6), 97–113 (1988)
23. T.N. Gia, T. Igor, V.K. Sarker, A.-M. Rahmani, T. Westerlund, P. Liljeberg, H. Tenhunen, IoT-based fall detection system with energy efficient sensor nodes, in *IEEE Nordic Circuits and Systems Conference (NORCAS'16)* (2016)
24. T.N. Gia, M. Jiang, A.-M. Rahmani, T. Westerlund, K. Mankodiya, P. Liljeberg, H. Tenhunen, Fog computing in body sensor networks: an energy efficient approach, in *IEEE International Body Sensor Networks Conference (BSN'15)* (2015)
25. T.N. Gia, M. Jiang, A.M. Rahmani, T. Westerlund, P. Liljeberg, H. Tenhunen, Fog computing in healthcare internet of things: A case study on ecg feature extraction, in *Proceeding of International Conference on Computer and Information Technology* (2015), pp. 356–363
26. B. Negash, A.M. Rahmani, T. Westerlund, P. Liljeberg, H. Tenhunen, Lisa: lightweight internet of things service bus architecture. Procedia Comput. Sci. **52**, 436–443 (2015). The 6th International Conference on Ambient Systems, Networks and Technologies (ANT-2015), the 5th International Conference on Sustainable Energy Information Technology (SEIT-2015)
27. N. Paul, T. Kohno, D.C. Klonoff, A review of the security of insulin pump infusion systems. J. Diabetes Sci. Technol. **5**(6), 1557–1562 (2011)
28. netfilter/iptables - nftables project. http://netfilter.org/projects/nftables/. Accessed 24 July 2015
29. G. Kambourakiset, E. Klaoudatou, S. Gritzalis, Securing medical sensor environments: the CodeBlue Framework case, in *Proceeding of the Second International Conference on Availability, Reliability and Security* (2007), pp. 637–643
30. R. Chakravorty, A programmable service architecture for mobile medical care, in *Proceeding of Fourth Annual IEEE International Conference on Pervasive Computing and Communications Workshops* (2006), pp. 5, 536

31. J. Ko, J.H. Lim, Y. Chen, R. Musvaloiu-E, A. Terzis, G.M. Masson, T. Gao, W. Destler, L. Selavo, R.P. Dutton, Medisn: medical emergency detection in sensor networks. ACM Trans. Embed. Comput. Syst. **10**(1), 11:1–11:29 (2010)
32. S.R. Moosavi, T.N. Gia, A. Rahmani, E. Nigussie, S. Virtanen, H. Tenhunen, J. Isoaho, SEA: a secure and efficient authentication and authorization architecture for IoT-based healthcare using smart gateways, in *Proceeding of 6th International Conference on Ambient Systems, Networks and Technologies* (2015), pp. 452–459
33. S.R. Moosavi, T.N. Gia, E. Nigussie, A. Rahmani, S. Virtanen, H. Tenhunen, J. Isoaho, Session resumption-based end-to-end security for healthcare internet-of-things, in *Proceeding of IEEE International Conference on Computer and Information Technology* (2015), pp. 581–588
34. PyCrypto API Documentation, https://pythonhosted.org/pycrypto/. Accessed 26 May 2016
35. Texas Instruments, Low-Power, 2-Channel, 24-Bit Analog Front-End for Biopotential Measurements (2012)
36. Texas Instruments, ECG Implementation on the TMS320C5515 DSP Medical Development Kit (MDK) with the ADS1298 ECG-FE (2011)
37. RTX Real-Time Operating System, http://www.keil.com/rl-arm/kernel.asp. 04 August 2015
38. R. Barry, *Using The FreeRTOS Real Time Kernel, Microchip PIC32 Edition*. FreeRTOS Tutorial Books (2010)
39. A. Dunkels, B. Gronvall, T. Voigt, Contiki - a lightweight and flexible operating system for tiny networked sensors, in *Proceeding of International Conference on Local Computer Networks* (2004), pp. 455–462
40. OMAP®4, PandaBoard System Reference Manual (2010), http://pandaboard.org. Accessed 04 August 2015
41. SmartRF06, Evaluation Board User's Guide (2013), http://www.ti.com/lit/ug/swru321a/swru321a.pdf. Accessed 04 August 2015
42. Olimex, MOD-ENC28J60 development board, Users Manual (2008). https://www.olimex.com/Products/Modules/Ethernet/MOD-ENC28J60/resources/MOD-ENC28J60.pdf. Accessed 04 August 2015
43. IEEE standard for medical device communication, overview and framework, in *ISO/IEEE 11073 Committee* (1996)
44. V.S. Miller et al., Data compression method. US4814746 A, Filing date August 11, 1986. Publication date March 21, 1989
45. G.B. Moody, R.G. Mark, The impact of the MIT-BIH arrhythmia database. Eng. Med. Biol. Mag. IEEE **20**(3), 45–50 (2001)
46. S. Luo, B. Ren, The monitoring and managing application of cloud computing based on internet of things. Comput. Methods Programs Biomed. **130**, 154–161 (2016)
47. J. Gomez, B. Oviedo, E. Zhuma, Patient monitoring system based on internet of things. Proc. Comput. Sci. **83**, 90–97 (2016)
48. G. Ji, W. Ouyang, K. Yang, G. Yang, Skin-attached sensor and artifact removal using cloud computing, in *7th International Conference on e-Health (eHealth2015)* (2015)
49. D. Bimschas, H. Hellbrück, R. Mietz, D. Pfisterer, K. Römer, T. Teubler, Middleware for smart gateways connecting sensornets to the Internet, in *Proceedings of the International Workshop on Middleware Tools, Services and Run-Time Support for Sensor Networks* (2010), pp. 8–14
50. Y. Shi, G. Ding, H. Wang, H.E. Roman, S. Lu, The fog computing service for healthcare, in *2015 2nd International Symposium on Future Information and Communication Technologies for Ubiquitous HealthCare (Ubi-HealthTech)*, May 2015, pp. 1–5
51. L. Prieto González, C. Jaedicke, J. Schubert, V. Stantchev, Fog computing architectures for healthcare: wireless performance and semantic opportunities. J. Inf. Commun. Ethics Soc. **14**(4), 334–349 (2016)
52. V. Stantchev, A. Barnawi, S. Ghulam, J. Schubert, G. Tamm, Smart items, fog and cloud computing as enablers of servitization in healthcare. Sens. Transducers **185**(2), 121 (2015)
53. Y. Cao, S. Chen, P. Hou, D. Brown, Fast: a fog computing assisted distributed analytics system to monitor fall for stroke mitigation, in *2015 IEEE International Conference on Networking, Architecture and Storage (NAS)* (IEEE, New York, 2015), pp. 2–11

54. A. Monteiro, H. Dubey, L. Mahler, Q. Yang, K. Mankodiya, Fit a fog computing device for speech teletreatments. arXiv preprint arXiv:1605.06236 (2016)
55. O. Fratu, C. Pena, R. Craciunescu, S. Halunga, Fog computing system for monitoring mild dementia and COPD patients-Romanian case study, in *2015 12th International Conference on Telecommunication in Modern Satellite, Cable and Broadcasting Services (TELSIKS)* (IEEE, New York, 2015), pp. 123–128
56. J.K. Zao, T.-T. Gan, C.-K. You, C.-E. Chung, Y.-T. Wang, S. José Rodríguez Méndez, T. Mullen, C. Yu, C. Kothe, C.-T. Hsiao, et al., Pervasive brain monitoring and data sharing based on multi-tier distributed computing and linked data technology. Front. Hum. Neurosci. **8**(370), 1–16 (2014)
57. M. Abu-Elkheir, H.S. Hassanein, S.M.A. Oteafy, Enhancing emergency response systems through leveraging crowdsensing and heterogeneous data, in *Wireless Communications and Mobile Computing Conference (IWCMC), 2016 International* (IEEE, New York, 2016), pp. 188–193

Index

A.M. Rahmani et al. (eds.), *Fog Computing in the Internet of Things*,
DOI 10.1007/978-3-319-57639-8

Printed by Printforce, the Netherlands